T0201778

Developing, Validating and Using Internal Ratings

Methodologies and Case Studies

Developing, Validating and Using Internal Ratings
Methodologies and Case Studies

Giacomo De Laurentis

Department of Finance and SDA Bocconi School of Management,
Bocconi University, Italy

Renato Maino

Lecturer, Bocconi University and Turin University, Italy

Luca Molteni

Department of Economics and SDA Bocconi School of Management,
Bocconi University, Italy

A John Wiley and Sons, Ltd., Publication

This edition first published 2010
© 2010 John Wiley & Sons Ltd.

Registered office
John Wiley & Sons Ltd, The Atrium, Southern Gate, Chichester, West Sussex, PO19 8SQ, United Kingdom

For details of our global editorial offices, for customer services and for information about how to apply for permission to reuse the copyright material in this book please see our website at www.wiley.com.

Library of Congress Cataloguing-in-Publication Data

De Laurentis, Giacomo
  Developing, validating, and using internal ratings : methodologies and case studies /
Giacomo De Laurentis, Renato Maino, Luca Molteni.
    p. cm.
  Includes bibliographical references and index.
  ISBN 978-0-470-71149-1 (cloth)
  1. Credit ratings. 2. Risk assessment. I. Maino, Renato. II. Molteni, Luca. III. Title.
  HG3751.5.D4 2010
  658.8′8 – dc22

                                                                          2010018735

A catalogue record for this book is available from the British Library.

ISBN 978-0-470-71149-1 (H/B)

Set in 10/12pt Times-Roman by Laserwords Private Limited, Chennai, India
Printed in Singapore by Markono Print Media Pte Ltd.

To Antonella, Daniela, Giuseppe, and Matteo

Contents

Preface

Banks are currently developing internal rating systems for both management and regulatory purposes. Model building, validation and use policies are key areas of research and/or implementation in banks, consultancy firms, and universities. They are extensively analyzed in this book, leveraging on international best practices as well as guidelines set by supervisory authorities. Two case studies are specifically devoted to building and validating statistical based models for borrower ratings.

This book starts by summarizing key concepts, measures and tools of credit risk management. Subsequently, it focuses on possible approaches to rating assignment, analyzing and comparing experts' judgment based approaches, statistical based models, heuristic and numerical tools. The first extensive case study follows. The model building process is described in detail, clarifying the main issues, how to use statistical tools and interpret results; univariate, bivariate, and multivariate stages of model building are discussed, highlighting the need to merge the knowledge of people with quantitative analysis skills with that of bank practitioners. Then validation processes are presented from various perspectives: internal and external (by supervisors), qualitative and quantitative, methodological and organizational. A second case study follows: a document for the internal validation unit, summarizing the process of building a shadow rating for assessing financial institutions creditworthiness, is proposed and analytically examined. Finally, conclusions are drawn: use policies are discussed in order to leverage on potentialities and managing limits of statistical based ratings.

The book is the result of academic research and the professional experience of its authors, mainly developed at the SDA Bocconi School of Management and Intesa Sanpaolo bank, as well as in consulting activities for many other financial institutions, including leasing and factoring companies. It focuses on quantitative tools, not forgetting that these tools cannot completely and uncritically substitute human judgment. Above all, in times of strong economic and financial discontinuities such as the period following the 2008 crisis, models and experience must be integrated and balanced out. This is why one of the fundamental tasks of this book is to merge different cultures, all of which are more and more necessary for modern banking:

- Statisticians must have good knowledge of the economic meaning of the data that they are working with and must realize the importance of human oversight in daily credit decisions.

- Credit and loan officers must have a fair understanding of the contents of quantitative tools, and properly understand how they can profit from their potentialities and what real limitations exist.

- Students attending credit risk management graduate and postgraduate courses must combine competences of finance, statistics and applicative tools, such as SAS and SPSS-PASW.

- Bank managers must set the optimal structure for lending processes and risk control processes, cleverly balancing competitive, management and regulatory needs.

As a consequence, the book tries to be useful to all and each of these groups of people and is structured as follows:

Chapter 1 introduces developments of credit risk management and recent insights gained from the financial crisis.

In Chapter 2, key concepts of credit risk management are summarized.

In Chapter 3, there is a description and a cross-examination of the main alternatives to rating assignment.

In Chapter 4, a case study based on real data is used to examine, step by step, the process of building and evaluating a statistical based borrower rating system for small and medium size enterprises aimed at being compliant with Basel II regulation. The data set is available on the book's website, www.wiley.com/go/validating. In the book, examples and syntax are based on the SPSS-PASW statistical package, which is powerful and friendly enough to be used both at universities and in business applications, whereas output and syntax files based on both SPSS-PASW and SAS are available on the book's website.

In Chapter 5, internal and regulatory validations of rating systems are discussed, considering both the qualitative and quantitative issues.

In Chapter 6, another case study is proposed, concerning the validation of a statistical based rating system for classifying financial institutions, in order to summarize some of the key tools of quantitative validation.

In Chapter 7, important issues related to organization and management profiles in the use of internal rating systems in banks' lending operations are discussed and conclusions are drawn.

Bibliography and a subject index complete the book.

In the book we refer to banks, but the term is used to indicate all financial institutions with lending activities.

The authors are pleased to acknowledge the great contributions of Nadeem Abbas, who has invaluably contributed to proof reading the entire book, and Daniele Tonini, who has reviewed some of the analyses in the book.

Giacomo De Laurentis
Renato Maino
Luca Molteni

About the authors

Giacomo De Laurentis, Full Professor of Banking and Finance at Bocconi University, Milan, Italy. Senior faculty member, SDA Bocconi School of Management. Director of Executive Education Open Programs Division, SDA Bocconi School of Management. Member of the Supervisory Body of McGraw-Hill and Standard & Poor's in Italy. Consultant to banks and member of domestic and international working groups on credit risk management and bank lending. In charge of credit risk management courses in the Master of Quantitative Finance and Credit Risk Management, other Masters of Science and Executive Masters at Bocconi University and SDA Bocconi School of Management.

Mail address: Università Bocconi, Department of Finance, Via Bocconi 8, 20136 Milano, Italy

Email address: giacomo.delaurentis@unibocconi.it

Renato Maino, Master in General Management at Insead. Member of international working groups on banking regulation, credit risk, liquidity risk. Intesa Sanpaolo Bank: former chief of Risk Capital & Policies, Risk Management Department; member of the Group's Financial Risk Committee; head of the Working Group for Rating Methodologies Development for Supervisory Recognition; head of the Working Group for Internal Capital Adequacy Assessment Process for Basel II. Arranger of international deals in corporate finance, structured finance and syndicated loans. Lecturer in risk management courses at Bocconi University, Milan, Italy, Politecnico of Turin, and University of Turin, Italy.

Mail address: via Rocciamelone 13, 10090 Villarbasse, Torino, Italy

Email address: renato.maino@unito.it

Luca Molteni, Assistant Professor of Statistics, Decision Sciences Department, Bocconi University, Milan, Italy. Senior faculty member, SDA Bocconi School of Management. CEO of Target Research (a market research and data mining consulting and services company). Consultant for risk management projects as an expert of risk management quantitative modelling.

Mail address: Università Bocconi, DEC Department, Via Roentgen 1, 20136 Milano, Italy

Email address: luca.molteni@unibocconi.it

1

The emergence of credit ratings tools

The 2008 financial crisis has shown that the reference context for supervisors, banks, public entities, non-financial firms, and even families had changed more than expected. From the perspective of banks' risk management, it is necessary to acknowledge the development of:

- New contracts (credit derivatives, loan sales, ABS, MBS, CDO, and so on).

- New tools to measure and manage risk (credit scoring, credit ratings, portfolio models, and the entire capital allocation framework).

- New players (hedge funds, sovereign funds, insurance companies, non-financial institutions entered into the financial arena).

- New regulations (Basel II, IAS/IFRS, etc.).

- New forces pushing towards profitability and growth (the apparently distant banking deregulation of the 1980s, contestable equity markets for banks and non-financial firms, management incentive schemes, etc.).

There are three key aspects to consider:

1. none of the aforementioned innovations can be considered relevant without the existence of the others;

2. each of the aforementioned innovations is useful to achieve higher levels of efficiency in managing banks;

3. all of these innovations are essentially procyclical.

Developing, Validating and Using Internal Ratings: Methodologies and Case Studies   Giacomo De Laurentis, Renato Maino and Luca Molteni   © 2010 John Wiley & Sons, Ltd

The problem is that the dynamic interaction among these innovations has created disequilibrium in both the financial and real economies.

As they are individually useful and all interconnected, a new equilibrium cannot be achieved by simply intervening in a few of them.

With this broader perspective in mind, we will focus on credit risk. In recent years, the conceptualization of credit risk has greatly improved. Concepts such as 'rating', 'expected loss', 'economic capital', and 'value at risk', just to name a few, have become familiar to bank managers. Applying these concepts has radically changed lending approaches in both commercial and investment banks, in fund management, in the insurance sector, and also for chief financial officers of large non-financial firms.

Changes concern tools, policies, organizational systems, and regulations related to underwriting, managing, and controlling credit risk. In particular, systems to measure expected losses (and their components: probability of default, loss given default, exposure at default) and unexpected losses (usually using portfolio VAR models) are tools which are nowadays regarded as a basic requirement. The competitive value of these tools pushes for an in-house building of models, also in accordance with the Basel Committee on Banking Supervision hopes.

The rating system is at the root of this revolution and represents the fundamental piece of every modern credit risk management system. According to the capital adequacy regulations, known as Basel II, the term rating system 'comprises all of the methods, processes, controls, and data collection and IT systems that support the assessment of credit risk, the assignment of internal risk ratings, and the quantification of default and loss estimates' (Basel Committee, 2004, p.394).

This signifies that 'risk models' must be part of a larger framework where, on one hand, their limits are perfectly understood and managed in order to avoid their dogmatic use, and, on the other hand, their formalization is not wasted by procedures characterized by excessive discretionary elements. To further outline this critical issue, how the current paradigm of risk measurements has been achieved in history and which decisions can be satisfactorily addressed by models (compared to those that should rest at the subjective discretion of managers) are addressed in this book.

The first provider of information concerning firms' creditworthiness was *Dun & Bradstreet*, which started in the beginning of the nineteenth century in the United States. At the end of the century, the first national financial market emerged in the United States; this financed immense infrastructures, such as railways connecting the east coast with the west coast. The issuing of bonds became widespread, in addition to more traditional shares. This evolution favored the creation of rating agencies, as they offer a systematic, autonomous, and independent judgment of bond quality. Since 1920, Moody's has produced ratings for more than 16 000 issuers and 30 000 issues; today it offers ratings for 4800 issuers. Standard & Poor's presently produces ratings of 3500 issuers. FITCH was created more recently by the merging of three other agencies: Fitch, IBCA, Duff & Phelps.

Internal ratings have a different anecdote. Banks started to internally classify borrowers in the United States in the second half of the 1980s when, after the collapse of more than 2800 savings banks, the FDIC and OCC introduced a formal subdivision of bank loans in different classes. The regulation required loans to be classified, with an initial confusion on what to rate (borrowers or facilities), in at least six classes, three of which today we would define as 'performing' and three as 'non-performing' (*substandard*, *doubtful* and *loss*). Provisions had to be set according to this classification of loans.

This regulatory innovation had an influential effect for banks, which started to classify counterparties and to accumulate statistical and financial data. During the 1990s, the most innovative banks were able to use a new analytical framework, based on the distinction of:

- the average frequency of default events for each rating class (the probability of default);

- the average loss in case of default (the loss given default);

- the amount involved in recovery processes for each facility (the exposure at default).

The new conceptual framework (initially adopted primarily by the investment banks, which are more involved in corporate finance) has rapidly shown its competitive value for commercial banks, in order to set more precise credit and commercial policies, and for defining pricing policies linked more to risk than to the mere bargaining power of the counterparts.

Quantitative data on borrowers and facilities' credit quality has allowed the creation of tools for portfolio analysis and for active asset management. Concepts such as diversification and capital at risk have been transposed to asset classes exposed to credit risk, and have enabled commercial banks to apply advanced and innovative forms of risk management.

By the end of the 1990s, after more than 10 years of positive experimentation, internal ratings appeared to be a good starting point for setting more risk-sensitive capital requirements for credit risk. The new regulation, known as Basel II, which has been gradually adopted by countries all over the world, has definitively consolidated these tools as essential measurements of credit risk, linking them with:

- The minimum capital requirement for credit risk, according to simplified representations of portfolios of loans (the First Pillar of the Basel II regulation).

- Capital requirements for concentration risk and the integration of credit risk with other risks (financial, operating, liquidity, business and strategy risks) in a holistic vision of capital adequacy (a key aspect of ICAAP, the *Internal Capital Adequacy Assessment Process* of the Second Pillar).

- Higher levels of disclosure of banks' exposure to risks in their communications to the market (the Third Pillar); this is functional to enhance the 'market discipline' by penalizing on financial markets those banks that take too much risk.

2

Classifications and key concepts of credit risk

2.1 Classification

2.1.1 Default mode and value-based valuations

Credit risk can be analyzed and measured from different perspectives. Table 2.1 shows a classification of diverse credit risk concepts. Each of the listed risks depends on specific circumstances. Default risk (also called counterparty risk, borrower risk and so forth, with minor differences in meaning) is an event related to the borrower's default. Recovery risk is related to the possibility that, in the event of default, the recovered amount is lower than the full amount due. Exposure risk is linked to the possible increase in the exposure at the time of default compared to the current exposure. A default-mode valuation (sometimes also referred to as 'loss-based valuation') considers all these three risks.

However, there are other relevant sources of potential losses over the loan's life. If we can sell assets exposed to credit risk (such as available-for-sale positions), we also have to take into account that the credit quality could possibly change over time and, consequently, the market value. Credit quality change is usually indicated by a rating migration; hence this risk is known as 'migration risk'.

In the new accounting principles (IAS 39), introduced in November 2009 by the International Accounting Standard Board (IASB), the amortized cost of financial instruments and impairment of 'loans and receivables' and of 'held-to-maturity positions' also depend on migration risk. Independently from the fact that 'true' negotiations occur, a periodic assessment of credit quality is required and, if

Developing, Validating and Using Internal Ratings: Methodologies and Case Studies   Giacomo De Laurentis, Renato Maino and Luca Molteni   © 2010 John Wiley & Sons, Ltd

Table 2.1   A classification of credit risk.

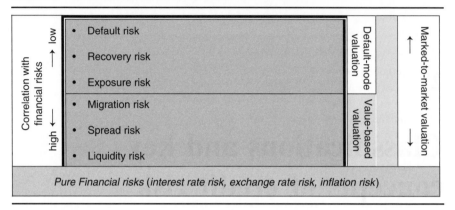

meaningful changes in credit quality arise, credit provisions have consequently to be arranged, and both losses and gains have to be recorded.

Finally, if positions exposed to credit risk are included in the trading book and valued at market prices, a new source of risk arises. In fact, even in the case of no rating migrations, investors may require different risk premiums due to different market conditions, devaluating or revaluating existing exposure values accordingly. This is the spread risk, and it generates losses and gains as well.

The recent financial crisis has underlined an additional risk (asset liquidity risk) related to the possibility that the market becomes less liquid and that credit exposures have to be sold, accepting lower values than expected (Finger, 2009a).

Credit ratings are critical tools for analyzing and measuring almost all these risk concepts. Consider for instance that risk premiums are usually rating sensitive, as well as market liquidity conditions.

2.1.2   Default risk

Without a counterparty's credit quality measure, in particular a default probability, we cannot pursue any modern credit risk management approach. The determination of this probability could be achieved through the following alternatives:

- The observation of historical default frequencies of borrowers' homogeneous classes. The borrowers' allocation to different credit quality classes has traditionally been based on subjective analysis, leveraging on analytic competences of skilled credit officers. Rating agencies have an almost secular track record of assigned ratings and default rates observed *ex post* per rating class.

- The use of mathematical and statistical tools, based on large databases. The bank's credit portfolios, which have thousands of positions observed in their historical behavior, allow the application of statistical methods. Models combine various types of information in a score that facilitates the borrowers' assignment to different risk classes. The same models permit a detailed *ex ante* measure of expected probability and facilitate monitoring over time.

- The combination of both judgmental and mechanical approaches (hybrid methods). Automatic classification is generated by statistical or numerical systems. Experts correct results by integrating qualitative aspects, in order to reach a classification that combines both potentialities (i.e., the systematic statistical analysis, expert competence and their ability to deal with soft information). Even in this case, the historical observation, combined with statistical methods, permits a default probability associated to each rating class to be reached.

- A completely different approach 'extracts' the implicit probability of default embedded in market prices (securities and stocks). The method can obviously only be applied to public listed counterparties on equity or securities markets.

The measure of default risk is the 'probability of default' within a specified time horizon, which is generally one year. However, it is also important to assess the cumulative probabilities when exposure extends beyond one year. The probability may be lower when considering shorter time horizons, but it never disappears. In overnight lending, too, we have a non-zero probability, given that sudden adverse events or 'hidden' situations to analysts may occur.

2.1.3 Recovery risk

The recovery rate is the complement to one of 'the loss in the event of default' (typically defined as LGD, Loss Given Default, expressed as a percentage). Note that here default is 'given', that is to say that it has already occurred.

In the event of default, the net position proceeds dependent on a series of elements. First of all, recovery procedures may be different according to the type of credit contracts involved the legal system and the court that has jurisdiction. The recovery rate also depends on the general economic conditions: results are better in periods of economic expansion. Defaulted borrowers' business sectors are important because assets values may be more or less volatile in different sectors. Also, covenants are important; these agreements between borrower and lender raise limits to borrower's actions, in order to provide some privileges to creditors. Some covenants, such as those limiting the disposal of important assets by the borrower, should be considered in LGD estimation. Other types of collateral may reduce the probability of default rather than the LGD; these are delicate aspects to models (Altman, Resti and Sironi, 2005; Moody's Investor Service, 2007).

Ex ante assessment of recovery rate (and corresponding loss given default) is by no means less complex than assessing the probability of default. Recovery rate data are much more difficult to collect, due to many reasons. Recoveries are often managed globally at the counterparty's position and, as a consequence, their reference to the original contracts, collaterals, and guarantees is often lost. Default files are mainly organized to comply with legal requirements, thus losing uniformity and comparability over time and across positions. Even when using the most sophisticated statistical techniques it is very difficult to build comprehensive models. Then, less sophisticated procedures are applied to these assessments, often adopting 'top down' procedures, which summarize the average LGD rates for a homogeneous set of facilities and guarantees. 'Loss given default ratings' (also known as 'severity ratings') are tools used to analyze and measure this risk.

2.1.4 Exposure risk

Exposure risk is defined as the amount of risk in the event of default. This amount is quite easily determined for term loans with a contractual reimbursement plan. The case of revolving credit lines whose balance depends more on external events and borrower's behavior is more complex. In this case, the due amount at default is typically calculated using model's specification, such as the following:

$$\text{Exposure at default} = \text{drawn} + (\text{limit} - \text{drawn})^* \text{LEQ}$$

where:

- drawn is the amount currently used (it can be zero in case of back-up lines, letters of credit, performance bonds or similar),

- limit is the maximum amount granted by the bank to the borrower for this credit facility,

- LEQ (Loan Equivalency Factor) is the rate of usage of the available limit, beyond the ordinary usage, in near-to-default situations.

In other cases, such as account receivables' financing, additional complexities originate from commercial events of non-compliance in contractual terms and conditions that can alter the amounts which are due from the buyer (the final debtor) to the bank. For derivative contracts, the due value in the event of default depends on market conditions of the underlying asset. The Exposure at Default (EAD) may therefore assume a probabilistic nature: its amount is a forecast of future events with an intrinsically stochastic approach. EAD models are the tools used to measure EAD risk.

2.2 Key concepts

2.2.1 Expected losses

A key concept of credit risk measurement is 'expected loss': it is the average loss generated in the long run by a group of credit facilities. The 'expected loss rate' is expressed as a percentage of the exposure at default.

The approach to determine expected loss may be financial or actuarial. In the former case, the loss is defined in terms of a decrease in market values resulting from any of the six credit risks listed in Table 2.1. In the latter case, the last three risks indicated in Table 2.1 (migration risk, spread risk, and liquidity risk) are not taken into consideration, only losses derived from the event of default are considered (therefore, it is generally known as 'default mode approach').

For banks, the expected loss is a sort of industrial cost that the lender has to face sooner or later. This cost is comparable to an insurance premium invested in mathematical risk-free reserves to cover losses over time (losses that actually fluctuate in different economic cycle phases).

Expected loss on a given time horizon is calculated by multiplying the following factors:

• probability of default

• severity of loss (LGD rate)

• exposure at default.

The expected loss rate, in percentage of EAD, only multiplies the first two measures.

2.2.2 Unexpected losses, VAR, and concentration risk

As the wording itself suggests, expected loss is expected (at least in the long term) and, therefore, it is a cost that is embedded into bank business and credit decisions. It is a sort of industrial cost of bank business. In short time horizons, banks' expected losses may strongly deviate from the long term average due to credit cycles and other events. Therefore, the most important risk lies in the fact that actual losses may deviate from expectations and, in particular, may become much higher than expected. In this case, the bank's capability to survive as a going concern is at stake. In short, the true concept of risk lies in unexpected loss rather than in expected loss.

Banks face unexpected losses by holding enough equity capital to absorb losses that are recorded in the income statement during bad times. Capital is replenished in good times by higher-than-expected profits. In credit risk management, capital has the fundamental role of absorbing unexpected losses and thus has to be commensurate with estimates of the loss variability over time.

In general, banks should hold enough capital to cover all risks, and not just credit risk. Bank managers must ensure they have an integrated view of risks in order to identify the appropriate level of capitalization. Calculating capital needs is only possible by using robust analytical risk models and measures. Credit risk measures are essential to contribute to a proper representation of risk.

From this perspective, ratings are key measures in determining credit contributions to the bank's overall risk. In fact, loss variability is very different for exposures in different rating classes. Therefore, on one hand, ratings directly produce measures of expected default rates and of expected loss given default, which impact credit provisions (costs written in banks' income statements). On the other hand, these measures help to differentiate exposures in terms of variability of default and LGD measures and their impact on banks' capital needs.

In many fields, unexpected losses are usually measured by standard deviation. However, in the case of credit risk, standard deviation is not an adequate measure of risk because the distribution (of losses, of default rates, and losses given default) is not symmetric (Figure 2.1).

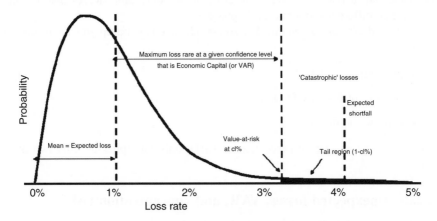

Figure 2.1   Loss rate distribution and economic capital.

In the case of credit risk, a better measure of variability is VAR (value at risk, here as a percentage of EAD), defined as the difference between the maximum loss rate at a certain confidence level and the expected loss rate, in a given time horizon. This measure of risk also indicates the amount of capital needed to protect the bank from failure at the stated level of confidence. This amount of capital is also known as 'economic capital'.

For instance, Figure 2.1 shows the maximum loss the portfolio might incur with a confidence level of cl% (say 99%, which means considering the worst loss rate in 99% of cases), the expected loss, and the value at risk. VAR defines the capital that must be put aside to overcome unexpected losses in 99% of the cases; the bank's insolvency is, therefore, confined to catastrophic loss rates whose probabilities are no more than one per cent $(1-cl\%)$[1].

In the case of credit risk, probability distributions are, by their nature, highly asymmetric. Adverse events may have a small probability but may impact significantly on banks' profit and loss accounts. The calculation of economic capital requires the identification of a probability density function. 'A credit risk

---

[1] The recent financial crisis has shown the opportunity to have measures on what may happen beyond the VAR threshold, in order to integrate VAR. Therefore, 'expected shortfall' is gaining consideration among risk managers; analytically, it is (in percentage) the average loss rate that is expected beyond a certain threshold defined in terms of confidence level.

model encompasses all of the policies, procedures and practices used by a bank in estimating a credit portfolio's probability density function' (Basel Committee, 1999a). In order to draw a loss (or default, LGD, EAD) distribution and calculate VAR measures, it is possible to adopt a parametric closed-form distribution, to use numerical simulations (such as Monte Carlo) or to use discrete probability solutions such as setting scenarios.

Up to now, expected losses and VAR measures (which are more specifically known as 'stand alone VAR') offer important summary measures of risk, but they do not take into account the risk deriving from portfolio concentration. The problem is that the sum of individual risks does not equal the portfolio risk. Increasing the number of loans in a portfolio and their diversification (in terms of borrowers, business sectors, regions, sizes and market segments, production technologies and so forth) reduces portfolio risk because of the less than perfect correlation among different exposures.

For this reason, a seventh risk concept should be added to Table 2.1 when considering the portfolio perspective: concentration risk. It arises in a credit portfolio where borrowers are exposed to common risk factors, that is, external conditions (interest rates, currencies, technological shifts and so forth). These risk factors may simultaneously impact on the willingness and ability to repay outstanding debts of a large number of counterparties. If the credit portfolio is specifically exposed to certain risk factors, the portfolio is 'concentrated' in respect to some external adverse events.

Traditionally, to avoid this risk, banks split claims on a large number of borrowers, limiting exposures and excessive market shares on individual customers. The idea was: the higher the portfolio granularity, the less risky the portfolio. In a context of quantitative credit risk management, the granularity criterion is integrated (and sometimes replaced) by the correlation analysis of events of default and of changes in credit exposures values.

'Full portfolio credit risk models' describe these diversification effects giving a measure of how much concentration is provided by the individual borrowers' risk factors; they also allow managing the credit portfolio risk profile as a whole or by segments. Without a credit portfolio model, it is not possible to analytically quantify the marginal risk attributable to different credit exposures, either if they are already underwritten or if they are just submitted for approval. Only if a portfolio model is available, is it then possible to estimate the concentration risk brought to the bank by each counterparty, transaction, facility type, market or commercial area. It is crucial to calculate default co-dependencies, that is to say, the possibility that more counterparts in the same risk scenario can jointly default or worsen their ratings.

There are two basic approaches to model default co-dependencies. The former is based on 'asset value correlation' and the framework proposed by Merton (1974): the effect of diversification lies in the possibility that the counterparties' value is influenced by external economic events. The event of joint default is related to the probability that two borrowers' assets values fall below their respective outstanding debt. The degree of diversification could therefore be measured by the correlation among assets values and by considering the outstanding debts of the two borrowers. The latter is based on a direct measure of the 'default correlation' in historical correlations of data of homogenous groups of borrowers (determined by elements such as business sector, size, geographical area of operation and so forth).

According to Markowitz's fundamental principle, only if the correlation coefficient is one is the portfolio risk equal to the sum of the individual borrowers' risks. On the contrary, as long as default events are not perfectly positively correlated, the bank will have to separately deal in different financial periods with its potential losses. Therefore, the bank can face the risk in a more orderly manner, with less intense fluctuations in provisioning and smaller committed bank capital.

In this perspective, it is also important to measure how individual exposures contribute to concentration risk, to the overall portfolio risk, and to the portfolio's economic capital. A 'marginal VAR' measure, indicating the additional credit portfolio risk implied by an individual exposure, is needed.

By defining:

- $UL_{portfolio}$ as the portfolio unexpected loss

- w_i as the weight of the ith loan on the overall portfolio

- $\rho_{i;portfolio}$ as the default correlation between the ith loan and overall portfolio

- ULC_i as the marginal contribution of the ith loan portfolio unexpected loss.

this marginal contribution can be expressed as:

$$ULC_i = \frac{\partial UL_{portfolio}}{\partial w_i} w_i$$

and in a traditional variance/covariance approach:

$$ULC_i = \rho_{i;portfolio} \, w_i \, UL_{portfolio}$$

ULC_i can be used in many useful calculations. For instance, a meaningful measure is given by the ith loan 'beta', defined as:

$$\beta_i = \frac{ULC_i/w_i}{UL_{portfolio}}$$

This measure compares the marginal i^{th} loan risk with the average risk at portfolio level. If β is larger than one, then the marginal risk adds more than the average risk to the portfolio; the reverse is true if β is lower than one. In this way, loans can be selected using betas, and thus it is possibly to immediately identify transactions that add concentration to the portfolio (i.e., they have a beta larger than one) and others that provide diversification benefits (beta smaller than one).

At different levels of the portfolio (individual loan, individual counterparty, counterparties' segments, sectors, markets and so forth), correlation coefficients ($\rho_{i;portfolio}$) and β_i can be calculated, achieving a quantitative measure of risk drivers. These measures can offer crucial information to set lending guidelines and to support credit relationship management. A number of publications, such as Resti and Sironi (2007) and De Servigny and Renault (2004), cover this content in more depth.

2.2.3   Risk adjusted pricing

Capital is costly because of shareholders' expectations on return on investment. Higher VARs indicate the need for higher economic capital; in turn, this implies the need for higher profits. Cost of capital multiplied by VAR is a lending cost, which has to be incorporated into credit spreads (if the bank is price setter) or considered as a cost (if the bank is price taker) in order to calculate risk adjusted performance measures. Lending decisions are as relevant for banks as investment decisions are for industrial companies; setting lending policies is as important to banks as selecting technology and business models for industrial companies.

The availability of information such as expected and unexpected losses can substantially innovate the way credit strategies are set. Today, the relevance of economic capital for pricing purposes is widely recognized (Saita, 2007). These measures must be incorporated into loan pricing. In theory, under the assumption of competitive financial markets, prices are exogenous to banks, which act as price takers and assess a deals expected return (*ex ante*) and actual return (*ex post*) by means of risk adjusted performance measures, such as the risk adjusted return on capital.

However, in practice, markets are segmented. For example, the loan market can be viewed as a mix of wholesale segments, where banks tend to behave more as price takers, and retail segments where, due to well known market imperfections (information asymmetries, monitoring costs and so forth), banks tend to set prices for their customers. In both cases, price may become a tool for credit policies and a way to shape the credit portfolio risk profile (in the medium term) by determining rules on how to combine risk and return of individual loans.

Therefore, the pricing policy drives loan underwriting and may incentivize cross-selling and customers' relationships management. At the bank's level, a risk-based pricing policy:

- structures the basis for active portfolio risk management (e.g., using credit derivatives);

- integrates credit risks with market risks and operational risks, supporting an effective economic capital budgeting;

- helps to formulate management objectives in terms of economic capital profitability at business units' level.

Many banks use risk adjusted performance measures to support pricing models; the most renowned is known as RAROC (risk adjusted return on capital) and has many variants, such as RARORAC (risk adjusted return on risk adjusted capital). In the late 1970s, the concept of RAROC was introduced for the first time by

Bankers Trust. This approach has become an integral part of the investment banks' valuations since the late 1980s (after the 1987 market crash and the 1991 credit crisis). Gradually, applications moved from management control (mainly at divisional level) to front line activities, in order to assess individual transactions. Since the mid 1990s, most of the major international transactions have been subject to prior verification of 'risk adjusted return' before loan marketing and underwriting.

The rationale of these applications is given by the theory of finance. The main assumption is that, ultimately, the value of different business lines depends on the ability to generate returns higher than those needed to reward the market risk premium required by capital which is absorbed to face risk. The Capital Asset Pricing Model (CAPM) provides a basis for defining the terms of the risk-return pattern. Broadly speaking and unless there are short-term deviations, credit must lie on the market risk/return line, taking into consideration correlation with other asset classes.

The credit spread has to be in proportion with the market risk premium, taking into consideration the risk premium of comparable investments. Otherwise (within the banking group or among different banks) market forces tend to align risk adjusted capital returns to the intrinsic value of underlying portfolios.

In particular, it is possible to fix the target return for the bank's credit risk-taking activities beyond the threshold of cost of capital. The best known practice is to establish a target level, for example, in terms of target Return on Equity (ROE; an accounting expression of the cost of equity) applied to the assets assigned to the division. The condition for value creation by a transaction is, therefore:

$$\text{RARORAC} > \text{ROE}_{\text{target}}$$

This relationship can also be expressed in terms of EVA (Economic Value Added):

$$\text{EVA} = (\text{RARORAC} - K_e) \times \text{Economic Capital}$$

in which K_e is the cost of shareholders' capital.

Risk-based pricing typically incorporates fundamental variables of a value-based management approach. For example, the pricing of credit products will include the cost of funding (such as an internal transfer rate on funds), the expected loss (in order to cover loan loss provisions), the allocated economic capital, and extra return (with respect to the cost of funding) as required by shareholders. Economic capital influences the credit process through the calculation of a (minimum) interest rate that is able to increase (or, at least, not decrease) shareholders' value. A simplified formula can be expressed as:

$$RARORAC = \frac{\text{Spread} + \text{Fees} - \text{Expected loss} - \text{Cost of capital} - \text{Cost of operations}}{\text{Economic capital}}$$

Depending on the product and the internal rules governing the credit process, decisions regarding prices can sometimes be overridden. For example, this situation could occur when considering the overall profitability of a specific customer's relationship or its desirability (due to reputational side-effects stemming from the customer relationship, even if it proves to be no longer economically profitable). Generally, these exceptions to the rule are strictly monitored and require the decision to be taken by a higher level of management.

Regardless of the role played by banks as price taker or price maker institutions, the process cannot be considered complete until feedback about the final outcome of the taken decision has been provided to management. The measurement of performance can be extended down to the customer level, through the analysis of customer profitability. Such an analysis aims to provide a broad and comprehensive view of all the costs, revenues, and risks (and, consequently, required economic capital) generated by each customer.

While implementation of this kind of analysis involves complex issues related to the aggregation of risks at the customer level, its use is evident in identifying unprofitable or marginally profitable customers who use resources (and above all capital) that could be allocated more efficiently to more profitable relationships.

This task is generally accomplished by segmenting customers in terms of ranges of (net) return per unit of risk. Provided the underlying inputs have been properly measured and allocated (not a simple task as it concerns risks and, even more, costs), this technique provides a straightforward indication of areas for intervention in managing customer profitability.

By providing evidence on the relative risk adjusted profitability of customer relationships (as well as products), economic capital can be used in optimizing the risk–return trade-off in bank portfolios. Recently, the adoption of these models has been accelerated because:

- investors are more sophisticated and promote the adoption of specific tools to maximize shareholders' value;

- banking groups are becoming large, complex, multinational conglomerates, and are more and more organized by distinct profit centers (business units). This implies an internal 'near-market' competition for resources and capital allocation. This organizational pattern requires risk adjusted performance measures and goals assigned throughout the whole structure.

In this context, ratings become not only a useful tool but also a necessary tool. In fact, without borrowers' creditworthiness measures, it is not possible to:

- operate on capital markets;

- manage the critical forces underlying value creation;

- compare the economic performance of business units or divisions and coordinate their actions.

3

Rating assignment methodologies

3.1 Introduction

In Chapter 2, the central role of ratings in supporting the new credit risk management architecture was emphasized. This role can be illustrated as an upside-down pyramid, with borrower's rating at its foundation (Figure 3.1). The event of default is one of the most significant source of losses in a banks profit and loss statements and assumes a central position in internal governance systems as well as in the eyes of specific supervisors' and monetary policy authorities' scrutiny.

Moreover, rating supports credit pricing and capital provisions to cover unexpected credit losses. These essential elements are at the foundation of many business decision making processes, touching all the organizational and operational aspects, up to business model selection, services offering, incentives and compensation systems, capital adequacy, internal controls systems, and internal checks and balances along the value chain of credit risk underwriting, management, and control.

Subsequently, the complex and delicate functions mentioned above pose relevant charges to rating assignment, far beyond only the technical requirement, even if it is considered a highly specialist component. Examined in the following chapters is how to calculate default probabilities through an appropriate rating system putting together coherent organizational processes, models, quantitative tools, and qualitative analyses.

Rating is an ordinal measure of the probability of the default event on a given time horizon, having the specific features of measurability, objectivity, and homogeneity, to properly confront counterparts and segments of the credit portfolio. Rating is the most important instrument that differentiates traditional from modern and quantitative credit risk management. The whole set of applications

Developing, Validating and Using Internal Ratings: Methodologies and Case Studies   Giacomo De Laurentis, Renato Maino and Luca Molteni   © 2010 John Wiley & Sons, Ltd

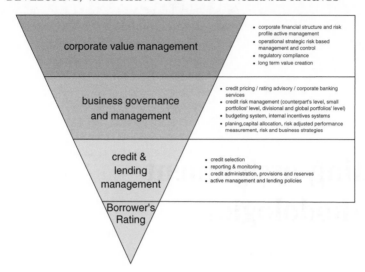

Figure 3.1   Credit governance system and borrower's rating.

mentioned before, which concerned expected losses, provisions, capital at risk, capital adequacy, risk adjusted performance measurement, credit pricing and control, and so forth, are essentially based on reliable probability measures.

Probabilities are expectations. If our *ex ante* assessment is accurate enough, over time, probabilities become actual observed frequencies, at least in pre-defined confidence intervals. This property implies that a specific organizational unit has to periodically verify any deviation out of confidence intervals, assessing impacts and effects, validating the assumptions and models that generated the *ex ante* expectations. The validation processes are returned to in Chapter 5.

As previously mentioned, the rating assessment backs up an important, well structured internal governance system, supporting decisions at the organization's different layers. This is why internal rating has to be as 'objective' as possible, in the sense that different teams of analysts – who are tackling the same circumstances, with the same level of information, applying the same methodology, in the same system of rules and procedures – have to arrive at a similar rating, accepting only minor misalignments. This is the only way to make decisions on a homogeneous, reliable and verifiable basis, maintaining full accountability over time.

Inevitably, there will be room for discretion, entrepreneurialism and subjectivity but a sound basis has to be provided to the whole process and control. This will not happen, obviously, if ratings were the result of individual and subjective analysis, contingently influenced by the point-in-time business environment or from highly personal competences that could be different each time, from one analyst to another.

These considerations do not imply that the credit analyst has to be substituted by tight procedures that stifle competences and professionalism. On the contrary, procedures have to put strong pressure on accountability and professionalism when needed to reach a better final decision. At the same time, it is necessary to avoid a sort of lenders' irresponsibility to fully take into account borrower's individual projects, initiatives, needs, and financial choices. In addition, lending decisions are not right or wrong; ratings only indicate that some choices are riskier than others, because a bank is responsible toward bondholders, depositors, and customers.

Therefore, rating systems have three desirable features in terms of measurability and verifiability, objectivity and homogeneity, and specificity:

- Measurability and verifiability: these mean that ratings have to give correct expectations in terms of default probabilities, adequately and continuously back tested.

- Objectivity and homogeneity: the former means that the rating system generates judgments only based on credit risk considerations, while avoiding any influence by other considerations; the latter means that ratings are comparable among portfolios, market segments, and customer types.

- Specificity: this means that the rating system is measuring the distance from the default event without any regards to other corporate financial features not directly related to it, such as short term fluctuations in stock prices.

These three features help to define a measure of appropriateness of internal rating systems and are decisive in depicting their distinctive suitability for credit management. However, the ability of different methodologies and approaches to deal with these desirable profiles is a matter of specific judgment, given the trade-offs existing among them.

Here, the following are distinguished and separately analyzed:

1. experts-based approaches

2. statistical-based models

3. heuristic and numerical approaches.

3.2 Experts-based approaches

3.2.1 Structured experts-based systems

Defaults are relatively rare events. Even in deeper recessions a default rate of around 2–5% is observed and each default appears like a highly individual story in approaching default, in recovery results, and in final outcome. A credit analyst (regardless of whether in a commercial bank or in an official rating agency) is, above all, an experienced person who is able to weigh intuitions and perceptions through the extensive knowledge accumulated in a long, devoted, and specialist career.

Also, economic theory regarding the framework of optimal corporate financial structure required a long development time, due to:

- lack of deep, homogenous, and reliable figures

- dominance of business and industrial competition problems rather than financial ones.

It is necessary to look back to the 1950s to see the first conceptual patterns on corporate financial matters, culminating with the Modigliani–Miller framework to corporate value and to the relevance of the financial structure. In the 1960s, starting from preliminary improvements in corporate finance stemming from Beaver (1966), the discipline became an independent, outstanding topic with an exponential amount of new research, knowledge, and empirical results. It is also necessary to have to recall the influential insight of Wilcox (1971), who applied 'gambler's ruin theory' to business failures using accounting data. Shortly after, from this perspective, the corporate financial problem was seen as a risky attempt to run the business by 'betting' the company's capital endowment. At the end of each round of betting, there would be a net cash in or net cash out. The 'company game' would end once the cash had finished. In formal terms, Wilcox proposed the relationship between:

- the probability of default ($P_{default}$),

- the probability of gains, m, and of losses $(1 - m)$; the constraint is that profits and losses must have the same magnitude,

- the company initial capital endowment, CN,

- the profit, U, for each round of the business game, in the form of

$$P_{default} = \left(\frac{1 - m}{m}\right)^{\frac{CN}{U}}$$

CN/U is the inverse of the return on equity ratio (ROE) and, in this approach, it is also the 'company's potential survival time'. Given the probability, m, then the process could be described in stochastic terms, identifying the range in which the company survival is assured or is going to experience the 'gambler's ruin'.

Many practical limitations impeded the model and, therefore, could not be applied in practice, confining it to a theoretical level. Nevertheless, the contribution was influential in the sense that:

- for the first time, an intrinsically probabilistic approach was applied to the corporate default description;

- the default event is embedded in the model, is not exogenously given, and stems from the company profile (profitability, capital, business turbulence and volatility);

- the explanatory variables are financial ones, linked with the business risk through the probability, m;

- there is the first definition of the 'time to default' concept, that has been used since the 1980s in the Poisson–Cox approach to credit risk.

These model features are very similar to Merton's model, which was proposed some years later; Merton's model is widely used today and is one of the most important innovations in credit risk management.

Another contribution that is worth mentioning is the 'point of no return theory', an expression that is common to war strategy or air navigation. The 'no return point' is the threshold beyond which one must continue on the current course of action, either because turning back is physically impossible or because, in doing so, it would be prohibitively expensive or dangerous. The theory is important because it has been defined using an intrinsically dynamic approach (Brunetti, Coda, and Favotto, 1984). The application to financial matters follows a very simple idea: the debt generates cash needs for interest payments and for the principal repayment at maturity. Cash is generated by the production, that is, by business and operations. If the production process is not generating enough cash, the company becomes insolvent. In mathematical terms this condition is defined as:

$$\frac{\partial EBIT}{\partial T} \geq \left(\frac{\partial (OF + \Delta D)}{\partial T} \right)$$

that is to say, the company will survive if the operational flow of funds (industrial margin plus net investments or divestments) is no less then interest charges and principal repayment, otherwise new debt is accumulated and the company is destined to fail. The balance between debt service and flow of funds from operations is consequently critical to achieve corporate financial sustainability over time. Therefore, the 'no return point' discriminates between sustainability and the potential path to default.

This idea plays an important role in credit quality analysis. Production, flow of funds, margins, and investments have to find a balance against financial costs; the default probability is, in some way, influenced by the 'safe margin', intended as the available cushion between operational cash generation and financial cash absorption. The company financial soundness is a function of the safe margins that the company is able to offer to lenders, like surpluses against failure to pay in sudden adverse conditions. It is an idea that is at the root of many frameworks of credit analysis, such as those used by rating agencies, and is at the basis of more structured approaches derived from Merton's option theory applied to corporate financial debt.

Credit quality analysis is historically concentrated on some sort of classification, with an aim to differentiate borrowers' default risk. Over time, the various tools changed from being mainly qualitative-based to being more quantitative-based. In the more structured approaches, the final judgment comes from a system of weights and indicators. Mostly, applied frameworks have symbolic acronyms such as:

- Four Cs: Character – Capital – Coverage – Collateral (proposed by Altman of New York University in various editions till the end of the 1990s).

- CAMELS: Capital adequacy – Asset quality – Management – Earnings – Liquidity – Sensitivity (J.P. Morgan approach).

- LAPS: Liquidity – Activity – Profitability – Structure (Goldman Sachs valuation system).

The final result is a class, that is, a discrete rank, not a probability. To reach a probability, an historical analysis has to be carried out, counting actual default frequencies observed per class over time.

During the 1980s and 1990s, industrial economics was deeply influenced by the competitive approach, proposed mainly by Porter (1980, 1985): economic phenomena, like innovation and globalization, deeply changed traditional financial analysis, creating the need to devote attention to the competitors' qualitative aspects, such as trading power, market position, and competitive advantages. These aspects had to be integrated with traditional quantitative aspects such as demand, costs, resources, and trading flows. Consequently, in the final judgment, it is critical to identify coherence, consistency, and appropriateness of the company's business conduct in relation to the business environment and competition.

Porter's important point is that qualitative features are as relevant as the financial structure and production capacity. Porter's publications can be considered today as at the roots of qualitative questionnaires and surveys that usually integrate the rating judgment, giving them solid theoretical grounds and conceptual references.

3.2.2 Agencies' ratings

The most relevant example of structured analysis applications is given by rating agencies (Ganguin and Bilardello, 2005). Their aim is to run a systematic survey on all determinants of default risk. There are a number of national and international rating agencies operating in all developed countries (Basel Committee, 2000b). The rating agencies' approach is very interesting because model-based and judgmental-based analyses are integrated (Adelson and Goldberg, 2009). They have the possibility to surmount the information asymmetry problem through a direct expert valuation, supported by information not accessible to other external valuators. Rating agencies' revenues derive for the most part by counterpart's fees; only a small amount is derived from the direct selling of economic information to investors and market participants. This business model is apparently very peculiar because of the obvious conflict of interests between the two parties. If the cost of the rating assignment is charged to companies that have the most benefit from it, how is it possible to be sure that this judgment will be reliable enough? Nevertheless, this business model is founded on a solid basis, as Nobel Laureate George Ackerloff and the 'lemon principle' can help us to understand. If there is a collective conviction among market participants that exchanged goods were of bad quality, the seller of better quality goods will encounter many difficulties in selling them, because they will have trouble in convincing people of the quality of his offer. In such circumstances, the seller of better quality goods:

- either tries to adapt, and switch to low quality goods in order to be aligned with the market judgment,

- or has to find a third party, a highly reputable expert, that could try to convince market participants that the offer is of really good quality and it is worth a higher price.

In the first case, the market will experience a suboptimal situation, because part of the potential offer (good quality products) will not be traded. In the second case, the market will benefit from the reliable external judgment, because of the opportunity to segment demand and to gain a wider number of negotiated goods.

Generally speaking, when there is information asymmetry among market participants (i.e., inability for market participants to have a complete and transparent evaluation of the quality of goods offered) only high reputation external appraisers can assure the quality of goods, overcoming the 'lemon' problem. Traders, investors, and buyers can lever on the expert judgment. Therefore, issuers are interested in demonstrating the credit quality of their issues, and rating agencies are interested in maintaining their reputation. The disruption in the evaluator's reputation is something that could induce a much wider market disruption (this observation is very important in light of the recent financial crisis, where rating agencies' structured-products judgments have been strongly criticized).

Consequently, the possibility to obtain privileged information of the counterparty's management visions, strategies, and budgeting is essential to a reliable rating agencies' business model; as a result, the structure of the rating process becomes a key part of the rating assignment process because it determines the possibility to have independent, objective, and sufficient insider information. Standard & Poor's (S&P) rating agency scheme is illustrated in Figure 3.2.

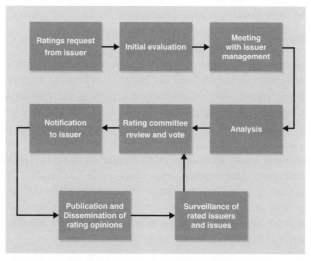

Figure 3.2   Decision making process for rating assignment at Standard & Poor's.
Source: *Standard & Poor's (2009c).*

Rating agencies' assignment methodologies are differentiated according to the counterparty's nature (corporations, countries, public entities and so forth) and/or according to the nature of products (structured finance, bonds and so forth). Here attention is concentrated on corporate borrowers. The final rating comes from two analytical areas (Figure 3.3): business risks and financial risks. This follows the fundamental distinction proposed by Modigliani and Miller in the 1950s.

Figure 3.3   Analytical areas for rating assignment at Standard & Poor's. Source: *Standard & Poor's (2008).*

The main financial ratios used by the Standard & Poor's rating agency are:

- profitability ratios from historical and projected operations, gross and net of taxes;

- coverage ratios such as cash flow from operations divided by interest and principal to be paid;

- quick and current liquidity ratios.

Generally speaking, the larger the cash flow margins from operations, the safer the financial structure; and, therefore, the better the borrower's credit rating. This general rule is integrated with considerations regarding the country of incorporation and/or of operations (so called 'sovereign' risk), the industry profile and the competitive environment, and the business sector (economic cycle sensitivity, profit and cash flow volatility, demand sustainability and so forth).

Other traditional analytical areas are: management's reputation, reliability, experience, and past performance; coherence and consistency in the firm's strategy; organization adequacy to competitive needs; diversifications in profit and cash flow sources; firm's resilience to business volatility and uncertainty. Recently, new analytical areas were introduced to take new sources of risk into account. The new analytical areas are as follows:

- internal governance quality (competence and integrity of board members and management, distribution and concentration of internal decision powers and layers, succession plans in case of critical management resources' resignation or vacation and so forth);

- environmental risks, technology and production processes compliance and sustainability;

- potential exposure to legal or institutional risks, and to main political events;

- potential hidden liabilities embedded, for instance, in workers' pension plans, health care, private assistance and insurance, bonuses, ESOP incentives and so forth.

Over time, aspects like internal governance, environmental compliance and liquidity have become crucial. Despite the effort in creating an objective basis for rating assignment, the agency rating 'is, in the end, an opinion. The rating experience is as much an art as it is a science' (Standard & Poor's, 1998, Foreword).

Under these considerations, it is worth noting that the rating process is very complex and is typically structured as follows: preliminary analysis, meetings with the counterparty under scrutiny, preparation of a rating dossier submitted by the Analytical Team to the Rating Committee (usually composed of 5–7 voting members), new detailed analysis if needed, final approval by the Rating Committee, official communication to and subsequent meeting with the counterparty, and, if necessary, a new approval process and rating submission to the Rating Committee. Moreover, the rating is not directly determined by ratios; for instance, the more favorable the business risk, the higher the financial leverage compatible with a given rating class (Table 3.1).

Table 3.1   Financial leverage (debt/capital, in percentage), business risk levels and ratings.

Company business risk profile / Rating category	AAA	AA	A	BBB	BB
Excellent	30	40	50	60	70
Above average	20	25	40	50	60
Average		15	30	40	55
Below average			25	35	45
Vulnerable				25	35

Source: Standard & Poor's (1998), page 19.

Generally speaking, favorable positions in some areas could be counterbalanced by less favorable positions in others, with some transformation criteria: financial ratios are not intended to be hurdle rates or prerequisites that should be achieved to attain a specific debt rating. Average ratios per rating class are *ex post* observations and not *ex ante* guidelines for rating assignment.

The rating industry has changed over time because of consolidation processes that have left only three big international players. It is worth noting that three competitors have different rating definitions. Moody's releases mainly issues ratings and far less issuers' ratings. On the contrary, S&P concentrates on providing a credit quality valuation referred to the issuer, despite the fact that the counterparty could be selectively insolvent on public listed bonds or on private liabilities. The company FITCH adopt an intermediate solution, offering an issuer rating, limited to the potential insolvency on publicly listed bonds, without considering the counterparty's private and commercial bank borrowings. Therefore, ratings released by the three international rating agencies are not directly comparable. This was clearly seen when, in the United Kingdom, British Railways defaulted and was privatized, while the outstanding debt was immediately covered by state guarantee. British Railways issues were set in 'selective default' by S&P while (coherently) having remained 'investment grade' for Moody's and 'speculative grade' for FITCH. In recent years, nonetheless, market pressure urged agencies to produce more comparable ratings, increasingly built on quantitative analyses, beyond qualitative ones, adopting a wider range of criteria. In particular, after the 'Corporate America scandals' (ENRON is probably the most renowned), new criteria were introduced, such as the so called 'Core earnings methodology' on treatment of stock options, multi annual revenues, derivatives and off-balance sheet exposures and so on. Liquidity profiles were also adopted to assess the short term liquidity position of firms, as well as the possibility to dismantle some investments or activities in case of severe recession and so forth. New corporate governance rules were also established with reference to conflict of interests, transparency, the quality of board members, investors' relations, minorities' rights protections and so on. Monitoring was enhanced and market signals (such as market prices on listed bonds and stocks) were taken into further consideration.

3.2.3   From borrower ratings to probabilities of default

The broad experience in rating assignment by agencies and the established methodology applied allow agencies to pile up a huge amount of empirical evidence about their judgments on predicted default rates. Until the 1990s, these data were only available for agencies' internal purposes; since then, these databases have also been sold to external researchers and became public throughout the credit analysts communities. Periodic publications followed, improving timeliness and specifications over time. Table 3.2 shows figures offered by Moody's rating agency on non-financial companies.

The basic principles at the foundation of these calculations are very straightforward:

- in the long run, given a homogenous population, actual frequencies converge to the central probability estimated, because of the law of large numbers (the average of the results obtained from a large number of trials should be close to the expected value, and will tend to become closer as more trials are performed);

- in the long run, if the population is homogeneous enough, actual frequencies are a good prediction of central probabilities.

In this perspective, when observations are averaged over time, probabilities could be inferred from the observation of average actual frequencies of default per rating class; these probabilities can be applied to infer the future of the population's behavior.

Table 3.2   Average cumulated annual default rates per issues cohorts, 1998/2007.

Initial rating	Average cumulated annual default rates at the end of each year (%)									
	year 1	year 2	year 3	year 4	year 5	year 6	year 7	year 8	year 9	year 10
Aaa	0.00	0.00	0.00	0.00	0.00	0.00	0.00	0.00	0.00	0.00
Aa1	0.00	0.00	0.00	0.00	0.00	0.00	0.00	0.00	0.00	0.00
Aa2	0.00	0.00	0.00	0.00	0.00	0.00	0.00	0.00	0.00	0.00
Aa3	0.00	0.00	0.00	0.00	0.00	0.06	0.17	0.17	0.17	0.17
A1	0.00	0.00	0.00	0.00	0.04	0.06	0.06	0.06	0.06	0.06
A2	0.05	0.11	0.25	0.35	0.46	0.52	0.52	0.52	0.52	0.52
A3	0.05	0.19	0.33	0.43	0.52	0.54	0.54	0.54	0.54	0.54
Baa1	0.21	0.49	0.76	0.90	0.95	1.04	1.26	1.58	1.66	1.66
Baa2	0.19	0.46	0.82	1.31	1.66	1.98	2.21	2.35	2.58	2.58
Baa3	0.39	0.93	1.54	2.21	3.00	3.42	3.85	4.33	4.49	4.49
Ba1	0.43	1.26	2.11	2.49	3.16	3.65	3.68	3.68	3.68	3.68
Ba2	0.77	1.71	2.81	4.03	4.78	5.06	5.45	6.48	7.53	10.16
Ba3	1.06	3.01	5.79	8.52	10.24	11.76	13.25	14.67	16.12	17.79
B1	1.71	5.76	10.21	14.07	17.14	19.59	21.21	23.75	26.61	28.37
B2	3.89	8.85	13.69	18.07	20.57	23.06	26.47	28.52	30.51	32.42
B3	6.18	13.24	21.02	27.63	33.35	39.09	42.57	45.19	48.76	51.11
Caa1	10.54	20.90	30.39	38.06	44.46	48.73	50.51	50.51	50.51	50.51
Caa2	18.98	29.51	37.24	42.71	44.99	46.83	46.83	46.83	46.83	46.83
Caa3	25.54	36.94	44.01	48.83	54.04	54.38	54.38	54.38	54.38	54.38
Ca-C	38.28	50.33	59.55	62.49	65.64	66.26	66.26	66.26	66.26	100.00
Investment Grade	0.10	0.25	0.43	0.61	0.77	0.88	0.99	1.08	1.13	1.13
Speculative Grade	4.69	9.27	13.70	17.28	19.79	21.77	23.27	24.64	26.04	27.38
All Rated	1.78	3.48	5.07	6.31	7.15	7.76	8.22	8.62	8.99	9.28

Source: Moody's (2008).

The availability of agencies' data also allows the calculation of the so-called migration frequencies, that is, the frequency of transition from one rating class to another; they offer an assessment of the 'migration risk', which has already been defined in the previous chapter. Tables 3.3 and 3.4 give examples of these migration matrices from Moody's publications: at the intersect of rows and columns there are relative frequencies of counterparties that have moved from the rating class indicated in each row to the rating class indicated in each column (as a percentage of the number of counterparties in the initial rating class). The acronym WR denotes 'withdrawn ratings', which are the ratings that have been removed for various reasons, only excluding default (to investigate this aspect further, see Gupton, Finger, and Batia, 1997, or de Servigny and Renault, 2004).

Table 3.3   One-year Moody's migration matrix (1970–2007 average).

		\multicolumn{9}{c}{Final rating class (%)}
		Aaa	Aa	A	Baa	Ba	B	Caa	Ca_C	Default	WR
Initial rating class	Aaa	89.1	7.1	0.6	0.0	0.0	0.0	0.0	0.0	0.0	3.2
	Aa	1.0	87.4	6.8	0.3	0.1	0.0	0.0	0.0	0.0	4.5
	A	0.1	2.7	87.5	4.9	0.5	0.1	0.0	0.0	0.0	4.1
	Baa	0.0	0.2	4.8	84.3	4.3	0.8	0.2	0.0	0.2	5.1
	Ba	0.0	0.1	0.4	5.7	75.7	7.7	0.5	0.0	1.1	8.8
	B	0.0	0.0	0.2	0.4	5.5	73.6	4.9	0.6	4.5	10.4
	Caa	0.0	0.0	0.0	0.2	0.7	9.9	58.1	3.6	14.7	12.8
	Ca-C	0.0	0.0	0.0	0.0	0.4	2.6	8.5	38.7	30.0	19.8

Source: Moody's (2008).

Table 3.4   Five-year Moody's migration matrix (1970–2007 average).

Cohort rating		\multicolumn{9}{c}{Final rating class (%)}
		Aaa	Aa	A	Baa	Ba	B	Caa	Ca_C	Default	WR
Initial rating class	Aaa	52.8	24.6	5.5	0.3	0.3	0.0	0.0	0.0	0.1	16.3
	Aa	3.3	50.4	21.7	3.3	0.6	0.2	0.0	0.0	0.2	20.3
	A	0.2	7.9	53.5	14.5	2.9	0.9	0.2	0.0	0.5	19.4
	Baa	0.2	1.3	13.7	46.9	9.4	3.0	0.5	0.1	1.8	23.2
	Ba	0.0	0.2	2.3	11.6	27.9	11.7	1.4	0.2	8.4	36.3
	B	0.0	0.1	0.3	1.5	7.2	21.8	4.5	0.7	22.4	41.5
	Caa	0.0	0.0	0.0	0.9	2.2	6.7	6.3	1.0	42.9	40.0
	Ca-C	0.0	0.0	0.0	0.0	0.3	2.3	1.5	2.5	47.1	46.3

Source: Moody's (2008).

The measures currently used by 'fixed income' market participants are based on:

- names: the number of issuers;
- Def: the number of names that have defaulted in the time horizon considered;
- PD: probability of default.

The default frequency in the horizon k, which is [t, (t+k)], is defined as:

$$PD_{\text{time horizon } k} = \frac{\text{Def}_t^{t+k}}{\text{Names}_t}$$

Given the sequence of default rates for a given issuers' class, the cumulated default frequency on horizon k is defined as:

$$PD_{\text{time horizon } k}^{\text{cumulated}} = \frac{\sum_{i=t}^{i=t+k} \text{Def}_i}{\text{Names}_t}$$

and the marginal default rate on the [t, (t+k)] horizon is defined as:

$$PD_k^{marg} = PD_{t+k}^{cumulated} - PD_t^{cumulated}$$

Finally, in regard to a future time horizon k, the 'forward probability' that is contingent to the survival rate at time t is defined as:

$$PD_{t;t+k}^{Forw} = \frac{(Def_{t+k} - Def_t)}{Names\ survived_t} = \frac{PD_{t+k}^{cumulated} - PD_t^{cumulated}}{1 - PD_t^{cumulated}}$$

Some relationships among these definitions can be examined further. The cumulated default rate $PD^{cumulated}$ may be calculated using forward probabilities (PD^{forw}) through the calculation of the 'forward survival rates' ($SR^{forw}_{t;\ t+k}$). These are the opposite (i.e., the complement to 1) of the PD^{forw}, and are as follows:

$$PD^{cumulated}{}_t = 1 - \left[(1 - PD^{forw}{}_1) \times (1 - PD^{forw}{}_2) \times (1 - PD^{forw}{}_3)\right.$$
$$\left. \times (1 - PD^{forw}{}_4) \times \dots \times (1 - PD^{forw}{}_n)\right]$$

If:

$$SR_{t;t+k}^{forw} = \left(1 - PD_{t;t+k}^{forw}\right)$$

then, the cumulated default rate can be expressed by the survival rates as:

$$PD_t^{cumulated} = 1 - \prod_{i=1}^{t} SR_i^{forw} \quad \text{and} \quad \left(1 - PD_t^{cumulated}\right) = \prod_{i=1}^{t} SR_i^{forw}$$

The 'annualized default rate' (ADR) can also be calculated. If it is necessary to price a credit risk exposed transaction on a five year time horizon, it is useful to reduce the five-year cumulated default rate to an annual basis for the purposes of calculation. The annualized default rate can be calculated by solving the following equation:

$$\left(1 - PD_t^{cumulated}\right) = \prod_{i=1}^{t} SR_i^{forw} = (1 - ADR_t)^t$$

Hence, the discrete time annualized default rate is:

$$ADR_t = 1 - \sqrt[t]{\prod_{i=1}^{t} SR_i^{forw}} = 1 - \sqrt[t]{\left(1 - PD_t^{cumulated}\right)}$$

Whereas, the continuous annual default rate is:

$$1 - PD_t^{cumulated} = e^{-ADR_t \times t}$$

and consequently:

$$ADR_t = -\frac{\ln\left(1 - PD_t^{cumulated}\right)}{t}$$

This formula gives the measure of a default rate, which is constant over time and generates the same cumulated default rate observed at the same maturity that was extracted from the empirical data.

Table 3.5 gives an example of the relationships between different measures that have been outlined above.

Table 3.5   Example of default frequencies for a given rating class.

	Years					
	1	2	3	4	5	
names$_{t=0}$	1000					
names$_t$	990	978	965	950	930	
default$_{cumulated;t}$	10	22	35	50	70	Formulas
PD^{cumulated}, %	1.00	2.20	3.50	5.00	7.00	$PD_k^{cumulated} = \dfrac{\sum_t^{t+k} Def_i}{Names_{t=0}}$
PD$^{marg}_k$, %	1.00	1.20	1.30	1.50	2.00	$PD_K^{marg} = PD_{t+k}^{cumulated} - PD_t^{cumulated}$
PD$^{forw}_k$, %	1.00	1.21	1.33	1.55	2.11	$PD_k^{Forward} = \dfrac{(Def_{t+k} - Def_t)}{Names\ survived_t}$
SR$^{cumul}_t$, %	99.00	97.80	96.50	95.00	93.00	$SR_t^{cumul} = \left(1 - PD_t^{cumulated}\right)$
SR$^{forw}_k$, %	99.00	98.80	98.70	98.50	98.00	$SR_k^{forw} = (1 - PD_k^{forw})$
ADR$_t$ discrete time, %	1.00	1.11	1.18	1.27	1.44	$ADR_t = 1 - \sqrt[t]{\left(1 - PD_t^{cumulated}\right)}$
ADR$_t$ continuous time, %	1.01	1.11	1.19	1.28	1.45	$ADR_t = -\dfrac{\ln\left(1 - PD_t^{cumulated}\right)}{t}$

With regard to the two final formulas, it must be borne in mind that they are shortcuts to solve pricing (and credit risk valuation) problems. In reality, it must be remembered that paths to default are not a steady and continuous process, but instead the paths to default present discontinuities and co-dependent events. Migrations are not 'Brownian random walks', but rather dependent and correlated transitions from one class to another over time. Moreover, credit quality and paths to default are managed both by counterparties and lenders. Actual observations prove that ratings become better than expected if the initial classes are low (bad) and, conversely, they become worse than expected if the initial classes are very high (good). These considerations have to be clearly taken into account when analyzing counterparties and markets, or when tackling matters such as defining the optimal corporate financial structure, performing a credit risk valuation or even measuring the risk of a credit portfolio.

Despite the fact that these methodologies are occasionally very complex and advanced, it is worth noting that default frequencies obviously have their limitations. They are influenced by the methodological choices of different rating agencies because:

- definitions are different through various rating agencies, so frequencies express dissimilar events;

- populations that generate observed frequencies are also different. As a matter of fact, many counterparts have only one or two official ratings, neglecting one or two of the other rating agencies;

- amounts rated are different, so when aggregated using weighted averages, the weights applied are dissimilar;

- initial rating for the same counterparts released by different rating agencies are not always similar.

Furthermore, official rating classes are an ordinal ranking, not a cardinal one: 'triple B' counterparty has a default propensity higher than a 'single A' and lower than a 'double B'. Actual default frequencies are only a proxy of this difference, not a rigorous statistical measure. Actual frequencies are only a surrogate of default probability. Over time, rating agencies have added more details to their publications and nowadays they also provide standard deviations of default rates observed over a long period. The variation coefficient, calculated using this data (standard deviation divided by mean), is really high through all the classes, mostly in the worst ones, which are highly influenced by the economic cycle. The distribution's fourth moment is high, showing a very large dispersion with fat tails and large probability of overlapping contiguous classes. Nevertheless, agencies' ratings are, *ex post*, the most performing measures among the available classifications for credit quality purposes.

3.2.4 Experts-based internal ratings used by banks

As previously mentioned, banks' internal classification methods have different backgrounds from agencies' ratings assignment processes. Nevertheless, sometimes their underlying processes are analogous; when banks adopt judgmental approaches to credit quality assessment, the data considered and the analytical processes are similar. Beyond any opinion on models' validity, rating agencies put forward a sound reference point to develop various internal analytical patterns. For many borrower segments, banks adopt more formalized (that is to say model based) approaches. Obviously, analytical solutions, weights, variables, components, and class granularities are different from one bank to another. But market competition and syndicated loans are strong forces leading to a higher convergence. In particular, where credit risk market prices are observable, banks tend to harmonize their valuation tools, favoring a substantial convergence of methods and results.

In principle, there is no proven inferiority or superiority of expert-based approaches versus formal ones, based on quantitative analysis such as statistical models. Certainly, judgment-based schemes need long lasting experience and repetitions, under a constant and consolidated method, to assure the convergence of judgments. It is very difficult to reach a consistency in this methodology and in its results because:

- organizational patterns are intrinsically dynamic, to adapt to changing market conditions and bank's growth, conditions that alter processes, procedures, customers' segments, organization appetite for risk and so forth;

- mergers and acquisitions that blend different credit portfolios, credit approval procedures, internal credit underwriting powers and so forth;

- over time, company culture will change, as well as experts' skills and analytical frameworks, in particular with reference to qualitative information.

Even if the predictive performances of these methods (read by appropriate accuracy measures) are good enough in a given period, it is not certain that the same performance will be reached in the future. This uncertainty could undermine the delicate and complex management systems that are based on internal rating systems in modern banking and financial organizations.

An assessment of the main features of expert-based rating systems along the three principles that have been previously introduced is proposed in Table 3.6.

Table 3.6   Summary of the main features of expert-based rating systems.

Criteria	Agencies' ratings	Internal experts-based rating systems
Measurability and verifiability	◕	◔
Objectivity and homogeneity	◕	◐
Specificity	●	◕

The circle is a measure of adequacy: full when completely compliant, empty if not compliant at all. Intermediate situations show different degrees of compliance.

3.3   Statistical-based models

3.3.1   Statistical-based classification

A quantitative model is a controlled description of certain real world mechanics. Models are used to explore the likely consequences of some assumptions under various circumstances. Models do not reveal the truth but are merely expressing points of view on how the world will probably behave. The distinctive features of a quantitative financial model are:

- the formal (quantitative) formulation, that explains the simplified view of the world we are trying to catch;

- the assumptions made to build the model, that set the foundation of relations among variables and the boundaries of the validity and scope of application of the model.

Generally speaking, the models which are employed in finance are based on simplifying assumptions about the phenomena that it is wished to predict; they should incorporate the vision of organizations' behavior, the possible economic events, and the reactions of market participants (that are probably using other models). Quantitative financial models therefore embody a mixture of statistics, behavioral psychology, and numerical methods.

Different assumptions and varying intended uses will, in general, lead to different models, and in finance, as in other domains, those intended for one use may not be suitable for other uses. Poor performance does not necessarily indicate defects in a model but rather that the model is used outside its specific realm. Consequently, it is necessary to define the type of models that are examined in the following pages very clearly. The models described in this book are mainly related to the assessment of default risk of unlisted firms, in effect, the risk that a counterparty may become insolvent in its obligations within a pre-defined time horizon. Generally speaking, this type of model is based on low frequency and non-publicly available data as well as mixed quantitative and qualitative variables. However, the methods proposed may also be useful in tackling the default risk assessment for large corporations, financial institutions, special purpose vehicles, government agencies, non-profit organizations, small businesses, and consumers. Also briefly touched upon are the credit risk valuation methods for listed financial and non-financial companies, briefly outlining the Merton approach.

The access to a wider range of quantitative information (mainly, but not only, from accounting and financial reports) pressured many researchers since the 1930s to try to generate classifications using statistical or numerical methodologies. In the mid 1930s, Fisher (1936) developed some preliminary applications. At the end of 1960s, Altman (1968) developed the first scoring methodology to classify corporate borrowers using discriminant analysis; this was a turning point for credit risk models.

In the following decades, corporate scoring systems were developed in more than 25 industrial and emerging countries. Scoring systems are part of credit risk management systems of many banks from western countries, as stated in a survey by the Basel Committee (2000b). To give an example of a non-Anglo-Saxon country, groundwork in this field was carried out in Italy during the late 1970s and early 1980s. In the early 1980s, the Financial Report Bureau was founded in Italy by more than 40 Italian banks; its objective was to collect and mutually distribute financial information about industrial Italian private and public limited companies. The availability of this database allowed the Bureau to develop a credit scoring model used by these banks during the 1980s and the 1990s. Nowadays, the use of quantitative methodologies is applied to the vast majority of borrowers and to ordinary lending decisions (De Laurentis, Saita and Sironi, 2004; Albareto *et al.*, 2008).

The first step in describing alternative models is to distinguish between structural approaches and reduced form approaches.

3.3.2   Structural approaches

Structural approaches are based on economic and financial theoretical assumptions describing the path to default. Model building is an estimate (similar to that of econometric models) of the formal relationships that associate the relevant variables of the theoretical model. This is opposite to the reduced form models, in which the final solution is reached using the most statistically suitable set of variables and disregarding the theoretical and conceptual causal relations among them.

This distinction became very apparent after the Merton (1974) proposal: default is seen as a technical event that occurs when the company's proprietary structure is no longer worthwhile. From the early 1980s, Merton's suggestion became widely used, creating a new foundation for credit risk measurements and analysis. According to this vision, cash flows out of a credit contract have the same structure as a European call option. In particular, the analogy implies the following:

- when the lender underwrites the contract, he is given the right to take possession of the borrower's assets if the borrower becomes insolvent;

- the lender sells a call option on the borrower's assets to the borrower, having the same maturity as debt, at the strike price equal to debt face value;

- at debt maturity, if the value of borrower's assets exceeds the debt face value, the borrower will pay the debt and shall retain full possession of his assets; otherwise, where the assets value is lower than the debt face value, the borrower has the convenience of missing debt payment and will resultantly lose possession of his assets.

This vision is very suggestive and offers workable solutions to overcome many analytical credit risk problems:

- By using a definition of default that is dependent on financial variables (market value of assets, debt amount, volatility of asset values) the Black Scholes Merton formula can be used to calculate default probability. There are five relevant variables: the debt face value (option strike), assets value (underlying option), maturity (option expiration date), assets value volatility (sigma), and market risk interest rate (for alternatives see Vasicek, 1984). This solution provides the probability that the option will be exercised, that is, the borrower will be insolvent.

- This result is intrinsically probabilistic. The option exercise depends on price movements and utility functions of the contract underwriters, hence from the simultaneous dynamics of the variables mentioned above. No other variables (such as macroeconomic conditions, legal constraints, jurisdictions, and default legal definitions) are implied in the default process.

- The default event is embedded in the model and is implicit in economic conditions at debt maturity.

- The default probability is not determined in a discrete space (as for agencies' ratings) but rather in a continuous space based on the stochastic dispersion of asset values against the default barrier, that is, the debt face value.

Merton's model is therefore a cause-and-effect approach: default prediction follows from input values. In this sense, Merton's model is a 'structural approach', because it provides analytical insight into the default process. Merton's insight offers many implications. The corporate equity is seen as a derivative contract, that is to say, the approach also offers a valuable methodology to the firm and its other liabilities. Moreover, as already stated, the option values for debt and equity implicitly include default probabilities to all horizons. This is a remarkable innovation in respect to the deductions of Modigliani and Miller. According to these two authors, the market value of the business ('assets') is equal to the market value of the fixed liabilities plus the market value of the equity; the firm financial structure is not related to the firm's value if default costs are negligible. The process of business management is devoted to maximizing the firm's value; the management of the financial structure is devoted to maximizing shareholders' value. No conflict ought to exist between these two objectives. In the Merton approach to the firm's financial structure, the equity is a call option on the market value of the assets. Thus, the value of the equity can be determined from the market value of the assets, the volatility of the assets, and the book value of the liabilities. That is to say, the business risk and the financial risk (assumed as independent risk by Modigliani and Miller) are simultaneously linked to one another by the firm's asset volatility. The firm's risk structure determines the optimal financial structure solution, which is based on the business risk profile and the state of the financial environment (interest rates, risk premium, equity and credit markets, investors' risk appetite).

More volatile businesses imply less debt/equity ratios and vice versa. The choice of the financial structure has an impact on the equity value because of the default probability (that is the probability that shareholders will lose their investments).

Following the Merton approach and applying Black Scholes Merton formula, the default probability is consequently given by:

$$PD = N\left(\frac{\ln(F) - \ln(V_A) - \mu T + \frac{1}{2}\sigma_A^2}{\sigma_A\sqrt{T}}\right)$$

where ln is the natural logarithm, F is the debt face value, V_a is the firm's asset value (equal to the market value of equity and net debt), μ is the 'risky world' expected return[1], T is the remaining time to maturity, σ_a is the instantaneous assets value volatility (standard deviation), N is the cumulated normal distribution operator.

The real world application of Merton's approach was neither easy nor direct. A firm's asset value and asset value volatility are both unobservable; the debt structure is usually complex, piling up many contracts, maturities, underlying guarantees, clauses and covenants. Black Scholes formula is highly simplified in many respects: interest rates, volatilities, and probability density functions of future events.

A practical solution was found in the early 1980s observing that the part in brackets of the mathematical expression described above is a standardized measure of the distance to the debt barrier, that is, the threshold beyond which the firm

---

[1] It is worth noting that the Black Scholes formula is valid in a risk neutral world. Hence the solution offers risk neutral probabilities. Here we are interested in real world (or so called 'physical') default probability because we are not interested in pricing debt but in describing actual defaults. To pass from actual to risk neutral default probabilities, a calculation is needed. The value of a credit contract can be defined as:

$$V_F = C_0 e^{-rt}(1 - q_t \omega)$$

in which V = credit market value of contract, C_t = initial credit face value, $\alpha\nu\delta\omega$ = loss given default. Using Black Scholes formula, it is possible to define:

$$q_{risk\ neutral\ world} = N(-Z)$$

$$p_{real\ world} = N(-Z')$$

in which

$$Z = \frac{\left[\ln\left(\frac{V}{D}\right) + r_{risk\ free} - \frac{\sigma^2}{2}\right]}{\sigma_v}; \quad Z' = \frac{\left[\ln\left(\frac{V}{D}\right) + r_{risky\ world} - \frac{\sigma^2}{2}\right]}{\sigma_v}$$

so:

$$q = N\left\{N^{-1}(p) + \frac{r_{risky\ world} - r_{risk\ free}}{\sigma_v}\right\}$$

But, from the Capital Asset Pricing Model theory, denoting the market risk premium as mrp:

$$r_{risky\ world} = r_{risk\ free} + \beta(mrp)$$

$$\beta = \frac{cov}{var_{mrp}}$$

then:

$$\frac{r_{risky\ world} - r_{risk\ free}}{\sigma_v} = \beta\frac{(mrp)}{\sigma_v} = \frac{cov}{\sigma_{mrp}^2} \times \frac{(mrp)}{\sigma_v} = \frac{cov}{\sigma_v \times \sigma_{mrp}} \times \frac{(mrp)}{\sigma_{mrp}} = \rho_{firm\ values;\ mrp}\lambda$$

in which λ is the Market Sharpe ratio. Subsequently, $q = N\{N^{-1}(p) + \rho\lambda\}$.

Upon deriving from market data C, r, t, LGD, p_t and ρ, it is possible to estimate λ on high frequency basis. The estimate of λ is quite stable over time (it suggests that spread variation is driven by PD variation, not risk premium). On the contrary, knowing λ, 'real world' default probability could be calculated starting from bond or CDS market prices.

goes into financial distress and default. This expression is then transformed into a probability by using the cumulated normal distribution function.

In addition, there is a relation (Itô, 1951) linking equity and asset value, based on their volatilities and the 'hedge ratio' of the Black Scholes formula. It has the form:

$$\sigma_{equity} E_0 = N(d_1) \sigma_{asset\ value} V_0$$

For listed companies on the stock market, equity values and equity volatilities are observable from market prices. Thus, this expression is absolutely relevant, as it allows calculation of asset values and asset values' volatility when equity market prices are known. A system of two unknown variables (V_a and σ_a) and two independent equations (the Black Scholes formula and Itô's lemma) has a simultaneous mathematical solution in the real numerical domain.

Therefore, the distance to default (DtD) can be calculated (assuming $T = 1$) as follows (Bohn, 2006):

$$DtD = \frac{\ln V_a - \ln F + \left(\mu_{risky} - \frac{\sigma_a^2}{2}\right) - \text{'other payouts'}}{\sigma_a} \cong \frac{\ln V - \ln F}{\sigma_a}$$

The default probability can be determined starting from this solution and using econometric calibration, even in continuous time, following the movements of equity prices and interest rates on the capital market. Despite the elegant solution offered by Merton's approach, in real world applications the solution is often reached by calibrating DtD on historical series of actual defaults. The KMV Company, established by McQuowm, Vasicek and Kealhofer, uses a statistical solution utilizing DtD as an explanatory variable to actual defaults. Solutions using statistical probability density functions (normal, lognormal or binomial) were found to be unreliable in real applications. These observations led Perraudin and Jackson (1999) to classify, also for regulatory purposes, Merton-based models as a scoring models.

DtD is a very powerful indicator that could also be used as an explicative variable in econometric or statistical models (mainly based on linear regression) to predict default probabilities.

These applications gained a huge success in the credit risk management world in order to rate publicly listed companies. Merton's approach is at the foundation of many credit trading platforms (supporting negotiations, arbitrage activities, and valuations) and tools for credit pricing, portfolio management, credit risk capital assessment, risk adjusted performance analysis, limit setting, and allocation strategies. Its main limit is that it is applicable only to liquid, publicly traded names. Also, in these cases, there is a continuous need for calibration; therefore, a specific maintenance is required. Small organizations cannot afford these analytical requirements, while nave approaches are highly unadvisable because of the great sensitivity of results to parameters and input measures.

Attempts were made to extend this approach to unlisted companies. Starting from the early 2000s, after some euphoria, these attempts were abandoned because of unavoidable obstacles. The main obstacles are:

- Prices are unobservable for unlisted companies. Using proxies or comparables, the methodology is very sensitive to some key parameters which causes results to become very unreliable.

- The use of comparables prices is not feasible when companies under analysis are medium sized enterprises. Comparables market prices become very scarce, smaller companies have very high specificity (their business is related to market niches or single production segments, largely idiosyncratic). In these circumstances, values and volatilities are very difficult to come across in a reliable way.

In comparison to agencies' ratings, Merton's approaches are:

- more sensitive to market movements and quicker and more accurate in describing the path to default;

- far more unstable (because of continuative movements in market prices, volatility, and interest rates). This aspect is not preferred by long term institutional investors that like to select investments based on counterparties' fundamentals, and dislike changing asset allocation too frequently.

Tarashev (2005) compared six different Merton-based structural credit risk models in order to evaluate the performance empirically, confronting the probabilities of default they deliver to *ex post* default rates. This paper found that theory-based default probabilities tend to closely match the actual level of credit risk and to account for its time path. At the same time, because of their high sensitivity, these models fail to fully reflect the dependence of credit risk on the business and credit cycles. Adding macro variables from the financial and real sides of the economy helps to substantially improve the forecasts of default rates.

3.3.3   Reduced form approaches

Reduced form models as opposed to structural models make no *ex ante* assumptions about the default causal drivers. The model's relationships are estimated in order to maximize the model's prediction power: firm characteristics are associated with default, using statistical methodologies to associate them to default data.

The default event is, therefore, exogenously given; it is a real life occurrence, a file in some bureau, a consequence of civil law, an official declaration of some lender, and/or a classification of some banks.

Independent variables are combined in models based on their contribution to the final result, that is, the default prediction on a pre-defined time horizon. The set of variables in the model can change their relevance in different stages of the economic cycle, in different sectors or for firms of different sizes. These are models without (or with a limited) economic theory, they are a sort of 'fishing expedition' in search of workable solutions.

The following is a practical example.

Suppose that the causal path to default is as follows: competitive gap → reduction in profitability margins → increase in working capital requirements (because of a higher increase in inventories and receivables than in payables) → banks' reluctance to lend more → liquidity shortage → insolvency and formal default.

A structural approach applied to listed companies could perceive this path as follows: reduction in profitability → reduction in equity price → more uncertainty in future profitability expectations → more volatility in equity prices → reduction in enterprise value and an increase in asset value volatility → banks' reluctance in granting new credit → stable debt barrier → sharp and progressive DtD reduction → gradual increase in default probability → early warning signals of credit quality deterioration and diffusion to credit prices → credit spread amplification → market perception of technical default situation.

As can be seen, a specific cause–effect process is clearly depicted.

What about a reduced form model? Assume that the default model is based on a function of four variables: return on sales, net working capital on sales, net financial leverage, and banks' short term debt divided by total debt. These variables are simultaneously observed (for instance, at year-end). No causal effect could therefore be perceived, because causes (competitive gap and return on sales erosion) are mixed with effects (net working capital and financial leverage increases). However, the model suggests that when such situations simultaneously occur, a default may occur soon after. If a company is able to manage these new working capital requirements by long-standing relationships with banks and new financial borrowings, it could overcome the potential crisis. However, it would be more difficult in a credit crunch period than in normal times.

In reduced form approaches there is a clear model risk: models intrinsically depend on the sample used to estimate them. Therefore, the possibility to generalize results requires a good degree of homogeneity between the development sample and the population to which the model will be applied. It should be clear at this point that different operational, business and organizational conditions, local market structures, fiscal and accounting regimes, contracts and applicable civil laws, may produce very different paths to default. As a consequence, this makes it clear that a model estimated in a given environment may be completely ineffective in another environment.

To give an idea of the relevance of this observation, consider a survey carried out at SanpaoloIMI Bank (Masera, 2001). A random sample of 1000 customers was extracted from the commercial lending portfolio. These companies were rated using the internal rating model and, at the same time, applying the last release of Altman's scoring formula (based on discriminant analysis, which will be extensively examined in the following paragraph). The purpose was to assess if:

- external models could be introduced in the banking organization without adaptation;

- variables proven to be relevant in external applications could also be useful to develop internal models, only making up coefficients and parameters.

The outcomes were very suggestive. When the classifications were compared, only 60% of good ratings were confirmed as good ratings by Altman's model, while 4% of these ratings were classified as being very close to defaulting by Altman's model. For bad rating classes, convergence was even lower or null. It could be considered that the problem was in model calibration and not in model structure: to overcome this objection, a new Altman-like model was developed, using the same variables but re-estimating parameters on the basis of the bank's internal sample. In this case, even if convergence was higher, results indicated that there was no chance to use the Altman-like model instead of the internal model. The main source of divergence was due to the role of variables that were highly country specific, such as working capital and liquidity ratios.

Therefore, these comparisons discourage the internal use of externally developed models: different market contexts require different analyses and models. This is not a trivial observation and there is, at the same time, a limitation and an opportunity in it:

- to develop an internal credit rating system is very demanding and requires much effort at both methodological and operational levels;

- to account on a reliable internal credit rating system is a value for the organization, a quality leap in valuation and analytical competence, a distinctive feature in competition.

When starting with a model building project, a strategic vision and a clear structural design at organizational level is required to adequately exploit benefits and advantages. The nature of the reduced form approaches impose the integration of statistics and quantitative methods with professional experience and qualitative information extracted from credit analysts from the initial stages of project development. In fact, even if these models are not based on relations expressing a causal theory of default, they have to be consistent with some theoretical economic expectation. For instance, if a profitability ratio is include in the model with a coefficient sign that indicates that the higher the ratio the higher default risk, we shall conclude that the model is not consistent with economic expectations.

Reduced form credit risk models could be classified into two main categories, statistical and numerical based. The latter comprises iterative algorithms that converge on a calibration useful to connect observed variables and actual defaults at some pre-defined minimum level of accuracy given the utility functions and performance measures. The former comprises models whose variables and relations are selected and calibrated by using statistical procedures.

These two approaches are the most modern and advanced. They are different from classifications based on the aggregation of various counterparts in homogeneous segments, defined by few counterparts' profiles (such as location, industry, sector, size, form of business incorporation, capitalization and so forth). These are referred to as 'top down' classifications because they segment counterparts based on their dominant profiles, without weighing variables and without combining them by specific algorithms; counterparts' profiles are typically categorical variables and they are used as a knife to split the portfolio into segments. Then, for each segment, the sample-based default rate will be used as an indicator of the probability of default for that segment of borrowers. Inversely, classification based on many variables whose values impact on results case by case are called 'bottom up'. Of course, there is a continuum between bottom up and top down approaches. Experts-based approaches are the most bottom up, but as they become more structured they reduce their capability of being case-specific. Numerical methods and statistical methods, even if highly mechanical, are considered as being 'quite' bottom up approaches because they take into account many variables characterizing the borrower (many of which are scale variables) and combine them by using specific algorithms.

An important family of statistical tools is usually referred to as scoring models; they are developed from quantitative and qualitative empirical data, determining appropriate variables and parameters to predict defaults. Today, linear discriminant analysis is still one of the most widely used statistical methods to estimate a scoring function.

3.3.4   Statistical methods: linear discriminant analysis

Models based on linear discriminant analysis (LDA) are reduced form models because the solution depends on the exogenous selection of variables, group composition, and the default definition (the event that divides the two groups of borrowers in the development sample). The performance of the model is determined by the ability of variables to give enough information to carry out the correct assignment of borrowers to the two groups of performing or defaulting cases.

The analysis produces a linear function of variables (known as 'scoring function'); variables are generally selected among a large set of accounting ratios, qualitative features, and judgments on the basis of their statistical significance (i.e., their contribution to the likelihood of default). Broadly speaking, the coefficients of the scoring functions represent the contributions (weights) of each ratio to the overall score. Scores are often referred to as Z-score or simply Z.

Once a good discriminant function has been estimated using historical data concerning performing and defaulted borrowers, it is possible to assign a new borrower to groups that were preliminarily defined (performing, defaulting) based on the score produced by the function. The number of discriminant functions generated by the solution of a LDA application is $(k-1)$, where k is the groups' number (in our case there are two, so there is one discriminant function).

Over time, the method has become more and more composite because of variegate developments; today there is a multitude of discriminant analysis methods. From here onwards, reference is mainly to the Ordinary Least Square method, which is the classic Fisher's linear discriminant analysis, analogous with the usual linear regression analysis. The method is based on a min–max optimization: to minimize variance inside the groups and maximize variance among groups.

Primarily, LDA has taxonomical purposes because it allows the initial population to be split into two groups which are more homogenous in terms of default probability, specifying an optimal discriminant Z-score threshold to distinguish between the two groups. Nevertheless, the scoring function can also be converted into a probabilistic measure, offering the distance from the average features of the two pre-defined groups, based on meaningful variables and proven to be relevant for discrimination.

The conceptual framework of reduced form models such as models based on LDA is summarized graphically in Figure 3.4. Let's assume that we are observing a population of firms during a given period. As time goes by, two groups emerge, one of insolvent (firms that fall into default) and the other of performing firms (no default has been filed in the considered time horizon: these are solvent firms). At the end of the period, there are two groups of well distinct firms: defaulted and performing firms. The problem is: given the firms' profile some time before the default (say t–k), is it possible to predict which firms will actually fall into default and which will not fall into default in the period between t-k and t?

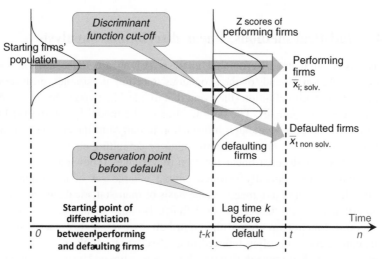

Figure 3.4   A simplified illustration of a LDA model framework applied to default prediction.

LDA assigns a Z-score to each firm at time t–k, on the basis of available (financial and non-financial) information concerning firms. In doing so, the groups of firms that at time t will be solvent or insolvent are indicated at time t–k by their Z-scores distributions. The differentiation between the two distributions is not perfect; in fact, given a Z cut-off, some firms that will become insolvent have a score similar to solvent firms, and vice versa. In other words, there is an overlapping between Z scores of performing and defaulting firms and, for a given cut-off, some firms are classified in the wrong area. These are the model's errors that are minimized by the statistical procedure used to estimate the scoring function.

LDA was one of the first statistical approaches used to solve the problem of attributing a counterpart to a credit quality class, starting from a set of quantitative attributes. Altman (1968) proposed a first model, which was based on a sample of 33 defaulted manufacturing firms and the same number of non-defaulted firms. For each firm, 22 financial ratios were available in the dataset. The estimated model included five discriminant variables and their optimal discriminant coefficients:

$$Z = 1.21x_1 + 1.40x_2 + 3.30x_3 + 0.6x_4 + 0.999x_5$$

where x_1 is working capital/total assets, x_2 is accrued capital reserves/total assets, x_3 is EBIT/total assets, x_4 is equity market value/face value of term debt, x_5 is sales/total assets, and Z is a number.

To understand the results, it is necessary to consider the fact that increasing Z implicates a more likely classification in the group of non-defaulted companies; therefore, all variables have coefficient signs aligned with financial theory. The discriminant threshold used to distinguish predicted defaulting from predicted performing companies is fixed at $Z = 2.675$ (also known as *cut-off value*).

A numerical example is shown in Table 3.7. Company ABC has a score of 3.19 and ought to be considered in a safety area. Leaving aside the independent variables' correlation (which is normally low) for the sake of simplicity, we can calculate the variables' contribution to the final result, as shown in Table 3.7.

We could also perform stress tests. For instance, if sales decrease by 10% and working capital requirements increase by 20% (typical consequence of a recession), the Z-score decreases to 2.77, which is closer to the cut-off point (meaning a higher probability of belonging to the default group). In these circumstances, the variables contribution for Company ABC will also change. For instance, the weight of working capital increases in the final result, changing from one-quarter to more than one-third; it gives, broadly speaking, a perception of elasticity to this crucial factor of credit quality. In particular, the new variables contribution to the rating of Company ABC will change as depicted in Table 3.8.

A recent application of LDA in the real world is the RiskCalc© model, developed by Moody's rating agency. It was specifically devoted to credit quality assessment of unlisted SMEs in different countries (Dwyer, Kocagil, and Stein, 2004). The model uses the usual financial information integrated by capital markets data, adopting a Merton approach as well. In this model (which is separately developed for many industrialized and emerging countries), the considered variables belong to different analytical areas, like profitability, financial leverage, debt coverage, growth, liquidity, assets, and size. To avoid over-fitting effects and in an attempt to have a complete view of the potential default determinants, the model is forced to use at least one variable per analytical area.

Table 3.7   Altman's Z-score calculation for Company ABC.

Asset & liabilities/equities			Profit & loss St.nt		
Fixed assets	100	*31.8%*	Sales	500 000	100.0%
Inventories	90	*28.7%*	EBITDA	35 000	7.0%
Receivables	120	*38.2%*	Net Financial Expenses	9 750	2.0%
Cash	4	*1.3%*	Taxes	8 333	1.7%
	314	**100%**	Profit	16 918	3.4%
Capital	80	*25.5%*	Dividends	11 335	2.3%
Accrued Capital reserves	40	*12.7%*	Accrued profits	5 583	1.1%
Financial debts	130	*41.4%*			
Payables	54	*17.2%*			
Other Net Liabilities	10	*3.2%*			
	314	**100%**			

Ratios for company ABC	(%)	Model coefficients	Ratio contributions for company ABC (%)
working capital / total assets	68	1.210	25.8
accrued capital reserves / total assets	13	1.400	5.6
EBIT / total assets	11	3.300	11.5
equity market value / face value of term debt	38	0.600	7.2
sales / total assets	159	0.999	49.9
Altman's Z-score		**3.191**	100

Table 3.8   New variables profile in a hypothetical recession for Company ABC.

Ratios for company ABC (%)		Model coefficients	Ratio contributions for company ABC (%)
working capital / total assets	72	1.210	31.4
accrued capital reserves / total assets	11	1.400	5.7
EBIT / total assets	8	3.300	10.0
equity market value / face value of term debt	34	0.600	7.3
sales / total assets,	126	0.999	45.6
Altman's Z-score		**2.768**	100.0

The model is estimated on a country-by-country basis. In the case of Italy, we have realized that the model takes evidence from the usual drivers of judgmental approaches:

- higher profitability and liquidity ratios have a substantially positive impact on credit quality, while higher financial leverage weakens financial robustness;

- growth has a double faceted role: when it is both very high and negative, the probability of default increases;

- activity ratios are equally multifaceted in their effects: huge inventories and high receivables lead to default, while investments (both tangible and intangible) either reduce the default probability or are not influential;

- company size is relevant because the larger ones are less prone to default.

LDA therefore optimizes variables coefficients to generate Z-scores that are able to minimize the 'overlapping zone' between performing and defaulting firms. The different variables help one another to determine a simultaneous solution of the variables weights. This approach allows the use of these models for a large variety of borrowers, in order to avoid developing different models for different businesses, as it happens when structuring expert based approaches.

3.3.4.1   Coefficient estimation in LDA

Assume that we have a dataset containing n observations (borrowers) described by q variables (each variable called x), split in two groups, respectively of performing and defaulted borrowers. The task is to find a discriminant function that enables us to assign a new borrower k, described in its x_k profile of q variables, to the performing (solvent) or defaulting (insolvent) groups, by maximizing a predefined measure of homogeneity (statistical proximity).

We can calculate variables means in each group, respectively defined in the two vectors $\overline{x_{solvent}}$ and $\overline{x_{insolvent}}$, known as groups' 'centroids'. The new observation k will then be assigned to either one or the other group on the basis of a minimization criterion, which is the following:

$$\min \left\{ \sum_{i=1}^{q} \left(x_{i;k} - \overline{x_{i;solv/insolv}} \right)^2 \right\}$$

or, in matrix algebra notation:

$$\min \left\{ \left(x_k - \overline{x_{solv/insolv}} \right)' \left(x_k - \overline{x_{solv/insolv}} \right) \right\}$$

This expression could be geometrically interpreted as the Euclidean distance of the new observation k to the two centroids (average profile of solvent and insolvent firms) in a q dimensions hyperspace. The lower the distance of k from one centroid, the closer the borrower k with that group, subject to the domain delimitated by the given q variables profile.

The q variables are obviously not independent to one another. They usually have interdependencies (correlation) that could duplicate meaningful information, biasing statistical estimates. To overcome this undesirable distortion, the Euclidean distance is transformed by taking these effects into consideration by the variables variance/covariance matrix. This criterion is the equivalent of using

Mahalanobis' 'generalized distance' (indicated by D). The k borrower attribution criterion becomes:

$$\min(D_k^2) = \min \left\{ \left(x_k - \overline{x_{\text{solv/insolv}}} \right)' \times C^{-1} \times \left(x_k - \overline{x_{\text{solv/insolv}}} \right) \right\}$$

where C is the q variables variance/covariance matrix considered in model development. The minimization of the function can be reached by estimating the Z-score function as:

$$Z_k = \sum_{j=1}^{n} \beta_j x_{k,j}$$

in which $\beta = (\overline{X}_{insolv} - \overline{X}_{solv})' C^{-1}$.

In this last formula, $\overline{X}_{insolv} - \overline{X}_{solv}$ denotes the difference between the centroids of the two groups. In other words, the goal of LDA is to find the combination of variables that:

- maximizes the homogeneity around the two centroids;

- minimizes the overlapping zone in which the two groups of borrowers are mixed and share similar Z-scores; in this area the model is wrongly classifying observations which have uncertain profiles.

We can calculate the Z values corresponding to the two centroids, respectively $\overline{Z}_{solv}$ and $\overline{Z}_{insolv}$, as the average Z for each group. Subject to certain conditions, it can be proved that the optimal discriminant threshold (cut-off point) is given by:

$$Z_{cut-off} = \frac{\overline{Z}_{solv} - \overline{Z}_{insolv}}{2}$$

In order to assign the borrower to one of the two groups, it is sufficient to compare Z_k of each k observation to the set Z_{cutoff}. The sign and size of all Z values are arbitrary; hence, the below/above threshold criterion could be reversed without any loss of generality and statistical meaning. Therefore, it is necessary to check each discriminant function one by one, to distinguish whether an increase in Z indicates higher or lower risk.

Applying LDA to a sample, a certain number of firms will be correctly classified in their solvent/insolvent groups; inevitably, some observations will be incorrectly classified in the opposite group. The aim of LDA is to minimize this incorrect classification according to an optimization criterion defined in statistical terms.

The result is a number (Z), not standardized and dimensionally dependent from the variables used; it indicates the distance on a linear axis (in the q variables hyperspace) between the two groups. The cut-off point is the optimal level of discrimination between the two groups; to simplify model's use and interpretation, sometimes it is set to zero by a very simple algebraic conversion.

Historically, these models were implemented to dichotomously distinguish between 'pass borrowers' (to grant loans to) and 'fail borrowers' (to avoid financing). Sometimes a gray area was considered, by placing two thresholds in order to have three ranges of Z-scores; the very safe borrowers, borrowers which need to be investigated further (possibly using credit analysts' expertise), and the very risky borrowers.

Today, we have two additional objectives: to assign ratings and to measure probability of default. These objectives are achieved by considering the score as an ascendant (descendant) grade of distance to the default, and categorizing scores in classes. This improvement does not yet satisfy the objective of obtaining a probability of default. To arrive at a probability measure, it is necessary to examine the concepts of model calibration and rating quantification.

LDA has some statistical requirements that should be met in order to avoid model inaccuracies and instability (Landau and Everitt, 2004; Giri, 2004; Stevens, 2002; Lyn, 2009), and are as follows:

 i. independent variables are normally distributed;

 ii. absence of heteroscedasticity, that is, the matrix C has to have similar values on the diagonal;

 iii. low independent variables multi-colinearity, that is, matrix C has to have homogenous and preferably low values off the diagonal, not statistically significant;

 iv. homogeneous independent variables variance around groups' centroids, that is, matrix C has to be (roughly) the same for firms in both solvent and insolvent groups.

The first three conditions can be overcome by adopting quadratic discriminant analysis instead of the linear discriminant analysis; in this case, we would use a model belonging to the group of Generalized Linear Models, which are will discuss later when considering logistic regression models. The fourth condition is a real life constraint because, as a matter of fact, insolvent firms typically have more prominent variances (as they have more diversified profiles) than solvent ones.

3.3.4.2   Model calibration and the cost of errors

Model calibration   In statistics, there are many uses of the term calibration. In its broader meaning, calibration is any type of fitting empirical data by a statistical model. For the Basel Committee (2005a, page 3) calibration is the quantification of the probability of default. In a more specific use, it indicates procedures to determine class membership probabilities of a given new observation. Here, calibration is referred to as the process of determining default probabilities for populations, starting from statistical based rating systems' outputs and taking into account the difference between development samples' default rates and populations' default rates. In other words, once the scoring function has been estimated and Z-scores have been obtained, there are still some steps to undertake before the model can actually be used. It is necessary to distinguish between the two cases.

In the first case: the model's task is to accept or reject credit applications (or even having a gray area classification), but multiple rating classes and an estimate of probability of default per rating class are not needed. In this case, model calibration simply consists of adjusting the Z-score cut-off in order to take into account differences in default rates of samples and of population. This circumstance is typical of applications of credit scoring models to consumer loan applications or it is simply an intermediate step in analyzing and validating models' performance.

In the second case: the model's task is to classify borrowers in different rating classes and to assign probabilities of default to borrowers. In this case, model calibration includes, in addition to cut-off adjustment, all steps for quantifying default probabilities starting from Z-score and, if needed, for rescaling them in order to take into account differences in default rates of samples and of population.

Model calibration: Z-score cut-off adjustment   In banks' loan portfolios, the number of defaulted firms is low compared to the number of non-defaulted firms. In randomly extracted samples, defaults are therefore very few in respects to performing firms. If this effect is not corrected when developing the model, the information on performing firms is overwhelming in comparison to the information on defaulted firms and, consequently, creates a bias in model estimation. In addition, LDA robustness suffers when variables have huge differences in their distribution in the two groups of borrowers. To limit these risks, the model building is carried out on more balanced samples, in which the two groups are more similar in size or, in extreme cases, have exactly the same sample size.

Therefore, when we apply model results to real populations, the risk is to over-predict defaults because, in the estimation sample, defaulted firms are over-represented. In other words, the frequency of borrowers classified as defaulting by the model is higher than the actual default rate in the population and, as a consequence, we need to calibrate results obtained from the development sample.

If a model based on discriminant analysis has not yet been quantified in order to associate the probability of default to scores or rating classes, and it is only used to classify borrowers to the two groups of performing and defaulting firms, calibration only leads to change the Z cut-off in order to achieve a frequency of borrowers classified as defaulting by the model equal to the default frequency in the actual population.

To calibrate a model based on discriminant analysis and used for classification purposes only, Bayes' theorem is applied. The theorem expresses the posterior probability (i.e., after evidence of scoring function variables values is observed) of a hypothesis (in our case, borrower's default), in terms of:

- the prior probabilities of the hypothesis, that is the probability of default when no evidence is collected on the specific borrower;

- the occurrence of evidence given the hypothesis, that is the probability of having a given Z-score in case the borrower defaults.

Consider that we have an ith borrower, described in its profile given by a variables vector X and summarized by a Z-score. Prior probabilities are identified as q and posterior probability as p. We can assume that:

- q_{insolv} and q_{solv} are the prior probabilities that the new ith observation will be attributed to the two groups without any regard to the information we have on them (the X vector); in our case $(q_{insolv} + q_{solv}) = 1$. Let's suppose that the default rate in real world population is 2.38%. If we lend money to a generic firm, having no other information, we could rationally suppose that q_{insolv} will be equal to 2.38% and q_{solv} will be equal to $(1 - 2.38\%) = 97.62\%$.

- The conditional probabilities to attribute the ith new observation, described in its profile X, respectively to the defaulted and performing groups are $p_{insolv}(X|insolv)$ and $p_{solv}(X|solv)$; they are generated by the model using a given sample. Suppose we have a perfectly balanced sample and the firm i is exactly on the cut-off point (hence the probability to be attributed to any of the two groups is 50%).

The simple probability (also called marginal probability) p(X) can be written as the sum of joint probabilities:

$$p(X) = q_{insolv} \cdot p_{insolv}(X|insolv) + q_{solv} \cdot p_{solv}(X|solv)$$

$$p(X) = 2.38\% \times 50\% + 97.62\% \times 50\% = 50\%$$

It is the probability of having the X profile of variables values (or its corresponding Z-score) in the considered sample, taking account of both defaulting and performing borrowers.

We are now in the position to use Bayes' theorem in order to adjust the cut-off by calibrating 'posterior probabilities'. The posterior probabilities, indicated by *p(insolv|X)* and *p(solv|X)*, are the probabilities that, given the evidence of the X variables, the firm i belongs to the group of defaulted or non-defaulted firms in the population. Using Bayes' formula:

$$p(insolv|X) = \frac{q_{insolv} \cdot p_{insolv}(X|insolv)}{p(X)}; \quad p(solv|X) = \frac{q_{solv} \cdot p_{solv}(X|solv)}{p(X)}$$

In our case, they will respectively be 2.38% and 97.62%.

In general, in order to calculate posterior probabilities, the framework in Table 3.9 can be used. Note that, in our case, the observation is located at the cut-off point of a balanced sample. Therefore, its conditional probability is 50%. When these circumstances are different, the conditional probabilities indicate in-the-sample probabilities of having a given value of Z-score for a solvent or insolvent firm. The sum of conditional probabilities is case specific and is not necessarily equal to 100%. The sum of joint probabilities represents the probability of having a given value of Z-score, considering both insolvent and solvent companies; again, this is a casespecific value, depending on assumptions.

Table 3.9   Bayes' theorem calculations.

Event	Prior probabilities (%)	Conditional probabilities of the ith observation (%)	Joint probabilities and their sum (%)	Posterior probabilities of the ith observation (%)
Default	q_{insolv} $= 2.38$	$P_{insolv}(X|insolv)$ $= 50$	$q_{insolv} \cdot p_{insolv}$ $(X|insolv)$ $= 1.19$	$p(insolv|X)$ $= 2.38$
Non-default	q_{solv} $= 97.62$	$p_{solv}(X|solv)$ $= 50$	$q_{solv} \cdot p_{solv}$ $(X|solv)$ $= 48.81$	$p(solv|X) =$ 97.62
Sum	100	–	$P(X) = 50$	100

The new unit i is assigned to the insolvent group if:

$$p(insolv|X) > p(solv|X)$$

Now consider firm i having a Z-score exactly equal to the cut-off point (for a model developed using balanced samples). Its Z-score would be 2.38%; as it is far less than 97.62%, the firm i has to be attributed to the performing group (and not to the group of defaulting firms). Therefore, the cut-off point has to be moved to take into consideration that the general population has a prior probability far less than we had in the sample.

To achieve a general formula, given Bayes' theorem and considering that p(X) is present in both items of $p(insolv|X) > p(solv|X)$, the formulation becomes:

$$q_{insolv} \cdot p_{insolv}(X|insolv) > q_{solv} \cdot p_{solv}(X|solv)$$

Hence, the relationship can be rewritten as:

$$\frac{p_{insolv}(X|insolv)}{p_{solv}(X|solv)} > \frac{q_{solv}}{q_{insolv}}$$

This formulation gives us the base to calibrate the correction to the cut-off point to tune results to the real world.

One of the LDA pre-requisites is that the distributions of the two groups are normal and similar. Given these conditions, Fisher's optimal solution for the cut-off point (obtained when *prior* chances to be attributed to any group is 50%) has to be relocated by the relation $\left[\ln \frac{q_{solv}}{q_{insolv}} \right]$. When the prior probabilities q_{insolv} and q_{solv} are equal (balanced sample), the relation is equal to zero, that is to say that no correction is needed to the cut-off point. If the population is not balanced, the cut-off point has to be moved by adding an amount given by the above relation to the original cut-off.

A numerical example can help. Assume we have a Z-score function, estimated using a perfectly balanced sample, and having a cut-off point at zero (for our convenience). As before, also assume that the total firms' population is made by all Italian borrowers (including non-financial corporations, family concerns, and small business) as recorded by the Italian Bank of Italy's Credit Register. During the last 30 years, the average default rate of this population (the a priori probability q_{insolv}) is 2.38%; the opposite (complement to one) is therefore 97.62% (q_{solv}, in our notation). The quantity to be added to the original Z-score cut-off is consequently:

$$\ln \frac{97.62\%}{2.38\%} = 3.71$$

The proportion of defaulted firms on the total population is called 'central tendency', a value that is of paramount importance in default probability estimation and in real life applications.

Cost of misclassification   A further important aspect is related to misclassifications and the cost of errors. No model is perfect when splitting the two groups of performing and defaulting firms. Hence, there will be borrowers that:

- are classified as potentially defaulted and would be rejected despite the fact that they will be solvent, therefore leading to the loss of business opportunities;

- are classified as potentially solvent and will be granted credit, but they will fall into default generating credit losses.

It is evident that the two types of errors are not equally costly when considering the potential loss arising from them. In the first case, the associated cost is an opportunity cost (regarding business lost, and usually calculated as the discounted net interest margin and fees not earned on rejected transactions), whereas the second case corresponds to the so-called loss given default examined in Chapter 2. For such reasons, it may be suitable to correct the cut-off point in order to take these different costs into consideration. Consider:

- $COST_{insolv/solv}$ the cost of false-performing firms that, once accepted, generate defaults and credit losses,

- $COST_{solv/insolv}$ the cost of false-defaulting firms, whose credit application rejection generate losses in business opportunities,

and assume that (hypothetically) $COST_{insolv/solv} = 60\%$ (current assessment of LGD) and $COST_{solv/insolv} = 15\%$ (net discounted values of business opportunities). The optimal cut-off point solution changes as follows:

$$\frac{p_{insolv}(X)}{p_{solv}(X)} > \frac{q_{solv} \times COST_{solv/insolv}}{q_{insolv} \times COST_{insolv/solv}}$$

Then, we have to add to the original cut-off an amount given by the relation $[\ln \frac{q_{solv} \times COST_{solv/insolv}}{q_{insolv} \times COST_{insolv/solv}}]$. Coming back to the previous example, by adding a weighted cost criterion, the cut-off point will be converted as follows:

$$\ln \frac{97.61\% \times 15\%}{2.38\% \times 60\%} = 2.33$$

This will be added to the original cut-off point to select new borrowers in pass/reject approaches, taking into consideration the cost of misclassification and the central default tendency of the population.

As a matter of fact, the sensitivity of the cut-off to the main variables (population default rate, misclassification costs and so forth) is very high. Moreover, moving the cut-off, the number rejected/accepted will generally change very intensively, determining different risk profiles of the credit portfolio originating from this choice. Cut-off relevance is so high that the responsibility to set *pro tempore* cut-offs is in the hand of offices different from those devoted to model building and credit analysis, and involves marketing and (often) planning departments. The cut-off point selection is driven by market trends, competitive position on various customer segments, past performances and budgets, overall credit portfolio profile, market risk environment (interest rates time structure, risk premium, capital market opportunities), funding costs and so forth in a holistic approach. These concepts are reviewed in Chapter 5.

3.3.4.3 From discriminant scores to default probabilities

LDA has the main function of giving a taxonomic classification of credit quality, given a set of pre-defined variables, splitting borrowers' transactions in potentially performing/defaulting firms. The typical decisions supported by LDA models are accept/reject ones.

Modern internal rating systems need something more than a binary decision, as they are based on the concept of default probability. So, if we want to use LDA techniques in this environment, we have to work out a probability, not only a classification in performing and defaulting firms' groups. We have to remember that a scoring function is not a probability but a distance expressed like a number (such as Euclidean distance, geometric distance – as the 'Mahalanobis distance' presented earlier – and so forth) that has a meaning in a domain of n dimensions hyperspace given by independent variables that describe borrowers to be classified. Therefore, when a LDA model's task is to classify borrowers in different rating classes and to assign probabilities of default to borrowers, model calibration includes, in addition to cut-off adjustment, all steps for quantifying default probabilities starting from Z-score and, if needed, for rescaling them in order to take into account differences in default rates of samples and of population.

The probability associated to the scoring function can be determined by adopting two main approaches: the first being empirical, the second analytical. The empirical approach is based on the observation of default rates associated to ascendant cumulative discrete percentiles of Z-scores in the sample. If the sample is large enough, a lot of scores are observed for defaulted and non-defaulted companies. We can then divide this distribution in discrete intervals. By calculating the default rate for each class of Z intervals, we can perceive the relationship between Z and default frequencies, which are our a priori probabilities of default. If the model is accurate and robust enough, default frequency is expected to move monotonically with Z values. Once the relationship between Z and default frequencies is set, we can infer that this relation will also hold in the future, extending these findings to new (out-of-sample) borrowers. Obviously, this correspondence has to be continuously monitored by periodic back testing to assess if the assumption is still holding.

The analytical approach is based again on the application of Bayes' theorem. Z-scores have no inferior or superior limits whereas probabilities range between zero and one. Let's denote again with p the posterior probabilities and with q the priors probabilities. Bayes' theorem states that:

$$p(insolv|X) = \frac{q_{insolv} \times p_{insolv}(X|insolv)}{q_{insolv} \times p_{insolv}(X|insolv) + q_{solv} \times p_{solv}(X|solv)}$$

The general function we want to achieve is a logistic function such as:

$$p(insolv|X) = \frac{1}{e^{\alpha+\beta X}}$$

It has the desired properties we are looking for: it ranges between zero and one and depends on Z-scores estimated by discriminant analysis.

In fact, it is possible to prove that, starting from the discriminant function Z, we obtain the above mentioned logistic expression by calculating:

$$\alpha = \ln\left(\frac{q_{solv}}{q_{insolv}}\right) - \frac{1}{2}(\overline{x_{solv}} - \overline{x_{insolv}})'C^{-1}(\overline{x_{solv}} - \overline{x_{insolv}})$$

$$= \ln\left(\frac{q_{solv}}{q_{insolv}}\right) - Z_{cut-off}$$

$$\beta = C^{-1}(\overline{x_{solv}} - \overline{x_{insolv}})$$

As previously mentioned, C is the variance/covariance matrix, and x are the vectors containing means of the two groups (defaulted and performing) in the set of variables X. Therefore, logistic transformation can be written as:

$$p(insolv|X) = \frac{1}{e^{\ln\left(\frac{q_{solv}}{q_{insolv}}\right)-z_{cut-off}+Z_i}}$$

in which $Z_{cut-off}$ is the cut-off point before calibration and $p(insolv|X)$ is the calibrated probability of default.

Once this calibration has been achieved, a calibration concerning cost misclassifications can also be applied.

3.3.5   Statistical methods: logistic regression

Logistic regression models (or LOGIT models) are a second group of statistical tools used to predict default. They are based on the analysis of dependency among variables. They belong to the family of Generalized Linear Models (GLMs), which are statistical models that represent an extension of classical linear models; both these families of models are used to analyze dependence, on average, of one or more dependent variables from one or more independent variables. GLMs are known as generalized because some fundamental statistical requirements of classical linear models, such as linear relations among independent and dependent variables or the constant variance of errors (homoscedasticity hypothesis), are relaxed. As such, GLMs allow modeling problems that would otherwise be impossible to manage by classical linear models.

A common characteristic of GLMs is the simultaneous presence of three elements:

1. *A Random Component:*   identifies the target variable and its probability function.

2. *A Systematic Component:*   specifies explanatory variables used in a linear predictor function.

3. *A Link Function:*   a function of the mean of the target variable that the model equates to the systematic component.

These three elements characterize linear regression models and are particularly useful when default risk is modeled.

Consider a random binary variable, which takes the value of one when a given event occurs (the borrower defaults), and otherwise it takes a value of zero. Define π as the probability that this event takes place; Y is a Bernoulli distribution with known characteristics (Y $\sim$ Ber(π), with π being the unknown parameter of the distribution). Y, as a Bernoullian distribution, has the following properties:

- $P(Y = 1) = \pi, \quad P(Y = 0) = 1 - \pi$
- $E(Y) = \pi, \quad Variance(Y) = \pi(1 - \pi)$
- $f(y : \pi) = \pi^y(1 - \pi)^{1-y} \quad per \ y \in \{0, 1\} \ e \ 0 \leq \pi \leq 1$

Therefore, Y is the random component of the model.

Now, consider a set of p variables $x_1, x_2, \ldots, x_p$ (with p lower than the number n of observations), $p + 1$ coefficients $\beta_0, \beta_1 \ldots, \beta_p$ and a function $g(\cdot)$. This function is known as the 'link function', which links variables x_j and their coefficients β_j with the expected value $E(Y_i) = \pi_i$ of the ith observation of Y, using a linear combination such as:

$$g(\pi_i) = \beta_0 + \beta_1 \cdot x_{i1} + \beta_2 \cdot x_{i2} +, \ldots, +\beta_p \cdot x_{ip} = \beta_0 + \sum_{j=1}^{p} \beta_j \cdot x_{ij} \quad i = 1, \ldots, n$$

This linear combination is known as a linear predictor of the model and is the systematic component of the model.

The link function $g(\pi_i)$ is monotonic and differentiable. It is possible to prove that it links the expected value of the dependent variable (the probability of default) with the systematic component of the model which consists of a linear combination of the explicative variables $x_1, x_2, \ldots, x_p$ and their effects β_i.

These effects are unknown and must be estimated. When a Bernoullian dependent variable is considered, it is possible to prove that:

$$g(\pi_i) = \log \frac{\pi_i}{1 - \pi_i} = \beta_0 + \sum_{j=1}^{p} \beta_j \cdot x_{ij} \qquad i = 1, \ldots, n$$

As a consequence, the link function is defined as the logarithm of the ratio between the probability of default and the probability of remaining a performing borrower. This ratio is known as 'odds' and, in this case, the link function $g(\cdot)$ is known as LOGIT (to say the logarithm of *odds*):

$$\text{logit}(\pi_i) = \log \frac{\pi_i}{1 - \pi_i}$$

The relation between the odds and the probability of default can be written as: $odds = \pi/(1 - \pi)$ or, alternatively, as $\pi = odds/(1 + odds)$.

Therefore, a LOGIT function associates the expected value of the dependent variable to a linear combination of the independent variables, which do not have any restrictive hypotheses. As a consequence, any type of explanatory variables is accepted (both quantitative and qualitative, and both scale and categorical), with no constraints concerning their distribution. The relationship between independent variables and the probability of default π is nonlinear (whereas the relation between logit (π) and independent variables is linear). To focus differences with the classical linear regression, consider that:

- In classical linear regression the dependent variable range is not limited and, therefore, may assume values outside the [0; 1] interval; when dealing with risk, this would be meaningless. Instead, a logarithmic relation has a dependent variable constrained between zero and one.

- The hypothesis of homoscedasticity of the classical linear model is meaningless in the case of a dichotomous dependent variable because, in this circumstance, variance is equal to $\pi(1 - \pi)$.

- The hypothesis testing of regression parameters is based on the assumptions that errors in prediction of the dependent variables are distributed similarly to normal curves. But, when the dependent variable only assumes values equal to zero or one, this assumption does not hold.

It is possible to prove that logit (π) can be rewritten in terms of default probability as:

$$\pi_i = \frac{1}{1 + e^{-\left(\beta_0 + \sum_{j=1}^{P} \beta_j x_{ij}\right)}} \qquad i = 1, \ldots, n$$

When there is only one explanatory variable x, the function can be graphically illustrated, as in Figure 3.5.

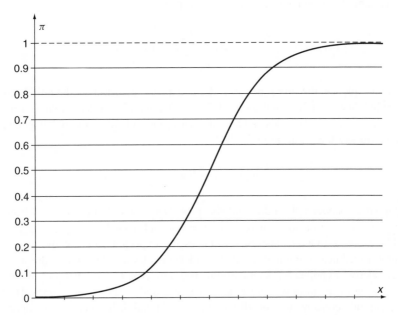

Figure 3.5   Default probability in the case of a single independent variable.

Note that this function is limited within the [0;1] interval, and is coherent with what we expect when examining a Bernoullian dependent variable: in this case, the coefficient β_1 sets the growth rate (negative or positive) of the curve and, if it is negative, the curve would decrease from one to zero; when β has a tendency towards zero, the curve flattens, and for $\beta = 0$ the dependent variable would be independent from the explanatory variable.

Now, let's clarify the meaning of 'odds'. As previously mentioned, they are the ratio between default probability and non-default probability.

Continuing to consider the case of having only one explanatory variable, the LOGIT function can be rewritten as:

$$\frac{\pi_i}{1 - \pi_i} = e^{(\beta_0 + \beta_1 x_{i1})} = e^{\beta_0} \cdot (e^{\beta_1})^{x_{i1}}$$

It is easy to interpret β. The odds are increased by a multiplicative factor e^β for one unit increase in x; in other words, odds for x + 1 equal odds for x multiplied by e^β. When $\beta = 0$, then $e^\beta = 1$ and thus odds do not change when x assumes different values, confirming what we have just mentioned regarding the case of independency. Therefore:

$$e^\beta = \frac{\text{odds after a unit change in the predictor}}{\text{original odds}}$$

We call this expression 'odds ratio'. Be cautious because the terminology used for odds is particularly confusing: often, the term that is used for odds is 'odds ratios' (and consequently this ratio should be defined as 'odds ratio ratio'!).

In logistic regression, coefficients are estimated by using the 'maximum likelihood estimation' (MLE) method; it selects the values of the model parameters that make data more likely than any other parameter' values would.

If the number of observations n is high enough, it is possible to derive asymptotic confidence intervals and hypothesis testing for the parameters. There are three methods to test the null hypothesis $H_0 : \beta_i = 0$ (indicating, as mentioned previously, that the probability of default is independent from explanatory variables). The most used method is the Wald statistic.

A final point needs to be clarified. Unlike LDA, logistic regression already yields sample-based estimates of the probability of default (PD), but this probability needs to be rescaled to the population's prior probability. Rescaling default probabilities is necessary when the proportion of bad borrowers in the sample is different from the actual composition of the portfolio (population) in which the logistic model has to be applied. The process of rescaling the results of logistic regression involves six steps (OeNB and FMA, 2004):

1. Calculation of the average default rate resulting from logistic regression using the development sample (π);

2. Conversion of this sample's average default rate into sample's average odds (SampleOdds), and calculated as follows:

$$Odds = \frac{\pi}{1 - \pi}$$

3. Calculation of the population's average default rate (prior probability of default) and conversion into population average odds (PopOdds);

4. Calculation of unscaled odds from default probability resulting from logistic regression for each borrower;

5. Multiplication of unscaled odds by the sample-specific scaling factor:

$$ScaledOdds = UnscaledOdds \cdot \frac{PopOdds}{SampleOdds}$$

6. Conversion of the resulting scaled odds into scaled default probabilities (π_S):

$$\pi_S = \frac{ScaledOdds}{1 + ScaledOdds}$$

This makes it possible to calculate a scaled default probability for each possible value resulting from logistic regression. Once these default probabilities have been assigned to grades in the rating scale, the calibration is complete.

It is important to assess the calibration of this prior-adjusted model. The population is stratified into quantiles, and the log odds mean is plotted against the log of default over performing rates in each quantile. In order to better reflect the population, default and performing rates are reweighted as described above for the population's prior probability. These weights are then used to create strata with equal total weights, and in calculating the mean odds and ratio of defaulting to performing. The population is divided among the maximum number of quantiles so that each contains at least one defaulting or performing case and so that the log odds are finite. For a perfectly calibrated model, the weighted mean predicted odds would equal the observed weighted odds for all strata, so the points would lie alongside the diagonal.

3.3.6   From partial ratings modules to the integrated model

Statistical models' independent variables may represent variegate types of information:

(a) firms' financial reports, summarized both by ratios and amounts;

(b) internal behavioral information, produced by operations and payments conveyed through the bank or deriving from periodical accounts balances, facility utilizations, and so on;

(c) external behavioral information, such as credit bureau reports, formal and informal notification about payments in arrears, dun letters, legal disputes and so on;

(d) credit register's behavioral data, summarizing a borrower's credit relationships with all reporting domestic banks financing it;

(e) qualitative assessments concerning firms' competitiveness, quality of management, judgments on strategies, plans, budgets, financial policies, supplier and customer relationships and so forth.

These sources of information are very different in many aspects: frequency, formalization, consistency, objectivity, statistical properties, and data type (scale, ordinal, nominal). Therefore, specific models are often built to separately manage each of these sources. These models are called 'modules' and produce specific scores based on the considered variables; they are then integrated into a final rating model, which is a 'second level model' that uses the modules' results as inputs to generate the final score. Each model represents a partial contribution to the identification of potential future defaults.

The advantages of using modules, rather than building a unitary one-level model, are:

- To facilitate models' usage and maintenance, separating modules using more dynamic data from modules which use more stable data. Internal behavioral data are the most dynamic (usually they are collected on a daily basis) and sensitive to the state of the economy, whereas qualitative information is seen as the steadiest because the firm's qualitative profile changes slowly, unless extraordinary events occur.

- To re-calculate only modules for which new data are available.

- To obtain a clear picture of the customer's profiles which are split into different analytical areas. Credit officers are facilitated to better understand the motivations and weaknesses of a firm's credit quality profile. At the same time, they can better assess the coherence and suitability of commercial proposals.

- All different areas of information contribute to the final rating; in one-level models the entire set of variables belonging to a specific area can be crowded out by other more powerful indicators.

- When a source of information is structurally unavailable (for instance, internal behavioral data for a prospective bank customer), different second-level models can be built by only using the available module, in order to tackle these circumstances.

- Information in each module has its peculiar statistical properties and, as a consequence, model building can be conveniently specialized.

Modules can be subsequently connected in parallel or in sequence, and some of them can be model based or rather judgment based. Figure 3.6 illustrates two possible solutions for the model structure. In Solution A (parallel approach), modules' outputs are the input for the final second-level rating model. In the example in Figure 3.6, judgment-based analysis is only added at the end of the process involving model-based modules; in other cases, judgment-based analysis can contribute to the final rating in parallel with other modules, as one of the modules which produces partial ratings. In Solution B there is an example of sequential approach (also known as the 'notching up/down approach'). Here, only financial information feeds the model whereas other modules notch financial model results up/down, by adopting structured approaches (notching tables or functions) or by involving experts into the notching process.

When modules are used in parallel, estimating the best function in order to consolidate them in a final rating model is not a simple task. On one hand, outputs from different datasets explain the same dependent variables; inevitably, these

• Solution A: Parallel approach.

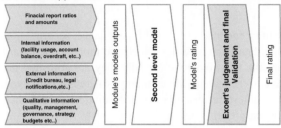

• Solution B: (sequential approach).

Figure 3.6   Possible architectures to structure rating modules in final rating.

outputs are correlated to each other and may lead to unstable and unreliable final results; specific tests have to be performed (such as the Durbin–Watson test). On the other hand, there are many possible methodological alternatives to be tested (most are discussed in this book) and important business considerations to be taken into account (mainly discussed in Chapter 7).

3.3.7   Unsupervised techniques for variance reduction and variables' association

Statistical approaches such as LDA and LOGIT methods are called 'supervised' because a dependent variable is defined (the default) and other independent variables are used to work out a reliable solution to give an *ex ante* prediction. Hereafter, we will illustrate other statistical techniques, defined as 'unsupervised' because a dependent variable is not explicitly defined. The borrowers or variables' sets are reduced, through simplifications and associations, in an optimal way, in order to obtain some sought-after features. Therefore, these statistical techniques are not directly aimed at forecasting potential defaults of borrowers but are useful in order to simplify available information. In particular, unsupervised statistical techniques are very useful for segmenting portfolios and for preliminary statistical explorations of borrowers' characteristics and variables' properties.

Given a database with observations in rows and variables in columns:

• 'Cluster analysis' operates in rows aggregating borrowers on the basis of their variables' profile. It leads to a sort of statistically-based top down segmentation of borrowers. Subsequently, the empirical default rate, calculated segment by segment, can be interpreted as the borrower' default probability of each segment. Cluster analysis can also be simply used as a preliminary exploration of borrowers characteristics.

- 'Principal component analysis', 'factor analysis', and 'canonical correlation analysis' all operate in columns in order to optimally transform the set of variables into a smaller one, which is statistically more significant.

In the future, these techniques may have a growing importance in order to build 'second generation' models in which the efficient use of information is essential.

3.3.7.1 Cluster analysis

The objective of cluster analysis is to explore if, in a dataset, groups of similar cases are observable. This classification is based on 'measures of distance' of observations' characteristics. Clusters of observations can be discovered using an aggregating criterion based on a specific homogeneity definition. Therefore, groups are subsets of observations that, in the statistical domain of the q variables, have some similarities due to analogous variables' profiles and are distinguishable from those belonging to other groups. The usefulness of clusters depends on:

- algorithms used to define them,

- economic meanings that we can find in the extracted aggregations.

Operationally, we can use two approaches: hierarchical or aggregative on the one hand, and partitioned or divisive on the other hand (Tan, Steinbach, and Kumar, 2006).

Hierarchical clustering   Hierarchical clustering creates a hierarchy of clusters, aggregating them on a case-by-case basis, to form a tree structure (often called *dendrogram*), with the leaves being clusters and the roots being the whole population. Algorithms for hierarchical clustering are generally agglomerative, in the sense that we start from the leaves and successively we merge clusters together, following branches till the roots. Given the choice of the linkage criterion, the pair-wise distances between observations are calculated by generating a table of distances. Then, the nearest cases are aggregated and each resulting aggregation is considered as a new unit. The process re-starts again, generating new aggregations, and so on until we reach the root. Cutting the tree at a given height determines the number and the size of clusters; often, a graph presentation is produced in order to immediately visualize the most convenient decision to make. Usually, the analysis produces:

- a small number of large clusters with high homogeneity,

- some small clusters with well defined and comprehensible specificities,

- single units not aggregated with others because of their high specificity.

Such a vision of data is of paramount importance for subsequent analytical activities, suggesting for instance to split groups that would be better analyzed by different models.

As mentioned before, the choice of the distance measure to use is crucial in order to have meaningful final results. The measures which are most used are:

- the Euclidean distance,

- the geometric distance (also called Mahalanobis distance), which takes into account different scales of data and correlations in the variables,

- the Hamming distance, which measures the minimum number of substitutions required to change one case into another,

- some homogeneity measures, such as the χ^2 test and the Fisher's F test.

Obviously, each criterion has its advantages and disadvantages. It is advisable to pre-treat variables in order to reach a similar magnitude and variability; indeed, many methods are highly influenced by variables' dimension and variance, and, thus, in order to avoid being unconsciously driven by some specific population feature, a preliminary transformation is highly recommended.

This method has many applications. One is the anomalies' detection; in the real world, many borrowers are outliers, that is to say, they have a very high specificity. In a bank's credit portfolio, start-ups, companies in liquidation procedures, and companies which have just merged or demerged, may have very different characteristics from other borrowers; in other cases, abnormalities could be a result of missing data and mistakes. Considering these cases while building models signifies biasing model coefficients estimates, diverting them from their central tendency. Cluster analysis offers a way to objectively identify these cases and to manage them separately from the remaining observations.

Divisive clustering   The partitional (or divisive) approach is the opposite of hierarchical clustering, because it starts at the root and recursively splits clusters by algorithms that assign each observation to the cluster whose center (also called centroid) is the nearest. The center is the average of all the points in the cluster. According to this approach, the number of clusters (k) is chosen exogenously using some rules. Then, k randomly generated clusters are determined with their cluster center. Each observation is assigned to the cluster whose center is the nearest; new cluster centers are re-calculated and the procedure is repeated until some convergence criterion is met. A typical criterion is that the cases assignment has not changed from one round to the next. At the end of the analysis a min−max solution is reached: the intra-group variance is minimized and the inter-group variance is maximized (subject to the constraint of the chosen clusters' number). Finally, the groups' profile is obtained showing the centroid and the variability around it. Some criteria help to avoid redundant iterations, avoiding useless or inefficient algorithm rounds.

The interpretation is the same for hierarchical methods: some groups are homogeneous and numerous while others are much less so, with other groups being typically residual with a small number of observations that are highly spread in the hyperspace which is defined by variables set. Compared to aggregative clustering, this approach could appear better as it tends to force our population in fewer groups, often aggregating hundreds of observations into some tens of clusters.

The disadvantage of these approaches is the required high calculation power. It exponentially increases with the number of initial observations and the number of iterative rounds of the algorithm. For such reasons, divisive applications are often limited to preliminary explorative analyses.

3.3.7.2   Principal component analysis and other similar methodologies

Let's return to our data table containing n cases described by q variables X. Using cluster analysis techniques we have dealt with the table by rows (cases). Now, we will examine the possibility to work on columns (variables). These modifications are aimed at substituting the q variables in a smaller (far smaller) number of new m variables, which are able to summarize the majority of the original total variance measured in the given q variables profiles. Moreover, this new m set that we will obtain has more desirable features, like orthogonality, less statistical 'noise' and analytical problems. Therefore, we can reach an efficient description, reducing the number of variables, linearly independent from each other.

Far beyond this perspective, these methods tend to unveil the database 'latent structure'. The assumption is that much of the phenomena are not immediately evident. In reality, the variables we identify and measure are only a part of the potential evidence of a complex, underlying phenomenon. A typical example is offered in the definition of intelligence in psychology. It is impossible to directly measure intelligence *per se*; the only method we have is to sum up the partial measures related to the different manifestations of intelligence in real life. Coming back to finance, for instance, firm profitability is something that is apparent at the conceptual level but, in reality, is only a composite measure (ROS, ROI, ROE, and so forth); nevertheless, we use the profitability concept as a mean to describe the probability of default; so we need good measures, possibly only one.

What can we do to reach this objective? The task is not only to identify some aspects of the firm's financial profile but also to define how many 'latent variables' are behind the ratio system. In other words, how much basic information do we have in a balance sheet that is useful for developing powerful models, avoiding redundancy but maintaining a sufficient comprehensiveness in describing the real circumstances?

Let's describe the first of these methods; one of the most well known is represented by 'principal components analysis'. With this statistical method, we aim to determine a transformation of the original $n \times q$ table into a second, derived, table $n \times w$, in which, for the generic j case (described through x_j in q original variables) the following relation holds:

$$w_j = X a_j$$

subject to the following conditions:

(a) Each w_i summarizes the maximum residual variance of the original q variables which are left unexplained by the $(i-1)$ previously extracted principal component. Obviously, the first one is the most general among all w extracted.

(b) Each w_i is perpendicular in respect to the others.

Regarding (a), we must introduce the concept of principal component communality. As mentioned before, each w has the property to summarize part of the variance of the original q variables. This performance (variance explained divided by total original variance) is called communality, and is expressed by the w_i principal component. The more general the component is (i.e., has high communality) the more relevant is the ability to summarize the original variables set in one new composed variable. This would compact information otherwise decomposed in many different features, measured by a plethora of figures.

Determination of the principal components is carried out by recursive algorithms. The method is begun by extracting the first component that reaches the maximum communality; then, the second is extracted by operating on the residuals which were not explained by the previous component, under the constraint of being orthogonal, until the entire original variables set is transformed into a new principal components set.

Doing this, because we are recursively trying to summarize as much as we can of the original total variance, the component that are extracted later contribute less to explain the original variables set. Starting from the first round, we could go on until we reach:

- a minimum pre-defined level of variance that we want to explain using the subset of new principal components,

- a minimum communality that assures us that we are compacting enough information when using the new component set, instead of the original variables set.

From a mathematical point of view, it is proven that the best first component is corresponding to the first eigenvalue (and associated eigenvector) of the variables set; the second corresponds to the first eigenvalue (and associated eigenvector) extracted on the residuals, and so on. The eigenvalue is also a measure of the corresponding communality associated to the extracted component. With this in mind, we can achieve a direct and easy rule. If the eigenvalue is more than one, we are sure that we are summarizing a part of the total variance that is more than the information given by an individual original variable (all the original variables, standardized, have contribution of one to the final variance). Conversely, if the eigenvalue is less than one, we are using a component that contributes less than an original variable to describe the original variability. Given this rule, a common practice is to only consider the principal component that has an eigenvalue of more than one.

If some original variables are not explained enough by the new principal component set, an iterative process can be performed. These variables are set apart from the database and a new principal component exercise is carried out until what could be summarized is compacted in the new principal component set and the remaining variables are used as they are. In this way, we can arrive at a very small number of features, some given by the new, orthogonally combined variables (principal components) and others by original ones.

Let's give an example to better understand these analytical opportunities. We can use results from a survey conducted in Italy on 52 firms based in northern Italy, which operate in the textile sector (Grassini, 2007). The goal of the survey was to find some aspects of sector competition; the variables refer to profitability performances, financial structure, liquidity, leverage, the firms positioning in the product/segment, R&D intensity, technological profile, and marketing organization. The variables list is shown in Table 3.10.

Table 3.11 shows the results of principal components extracted, that is to say, the transformation of the original variables set in another set with desirable statistical features. The new variables (components) are orthogonal and (as in a waterfall) explain the original variance in descending order.

The first three components summarize around 81% of the total original variance, and eigenvalues explain how much variance is accounted by each component. The first component, despite being the most effective, takes into account 40% of the total. Therefore, a model based only on one component does not account for more than this and would be too inefficient. By adding two other features (represented by the other two components), we can obtain a picture of four-fifths of the total variability, which can be considered as a good success. Table 3.12 shows the correlation coefficients between the original variables set and the first three components. This table is essential for detecting the meaning of the new variables (components) and, therefore, to understand them carefully.

The first component is the feature that characterizes the variables set the most. In this case, we can see that it is highly characterized by the liquidity variables, either directly (for current liquidity and quick liquidity ratios) or inversely correlated (financial leverage). A good liquidity structure reduces leverage and vice versa; so the sign and size of the relationships are as expected. There are minor (but not marginal) effects on operational and shareholders' profitability: that is, liquidity also contributes to boost firm's performances; this relationship is also supported by results of the Harvard Business School's Profit Impact of Market Strategy (PIMS) database long term analysis (Buzzell and Gale, 1987, 2004).

The second component focuses on profitability. The lighter the capital intensity of production, the better the generated results are, particularly in respect of working capital requirements.

Table 3.10   Variables, statistical profile and correlation matrix.

Variables typology	Variable denomination	Definition
Profitability performance	ROE	net profit/net shareholders capital
	ROI	EBIT/invested capital
	SHARE	market share (in %)
Financial structure on short and medium term horizon	CR	current assets/Current liabilities
	QR	liquidity/current liabilities
	MTCI	(current liabilities + permanent liabilities)/invested capital
Intangibles (royalties, R&D expenses, product development and marketing)	R&S	Intangibles fixed assets/ invested capital (in percentage)

Ratios	Mean	Minimum value	Maximum value	Standard deviation	Variability coefficient	Asymmetry	Kurtosis
ROE	0.067	−0.279	0.688	0.174	2.595	1.649	5.088
ROI	0.076	−0.012	0.412	0.078	1.024	2.240	5.985
CR	1.309	0.685	3.212	0.495	0.378	1.959	4.564
QR	0.884	0.169	2.256	0.409	0.463	1.597	2.896
MTCI	0.787	0.360	1.034	0.151	0.192	−0.976	0.724
SHARE (%)	0.903	0.016	6.235	1.258	1.393	2.594	7.076
R&S (%)	0.883	0.004	6.120	1.128	1.277	2.756	9.625

Ratios	ROE	ROI	CR	QR	MTCI	SHARE (%)	R&S(%)
ROE	1.000						
ROI	**0.830**	1.000					
CR	−0.002	0.068	1.000				
QR	0.034	0.193	**0.871**	1.000			
MTCI	−0.181	−0.333	**−0.782**	**−0.749**	1.000		
SHARE (%)	0.086	0.117	−0.128	−0.059	0.002	1.000	
R&S (%)	−0.265	−0.144	−0.155	−0.094	−0.013	0.086	1.000

Bold: statistically meaningful correlation.

The third component summarizes the effects of intangibles, market share and R&D investments. In fact, R&D and intangibles are related to the firm's market share, that is to say, to the firm's size. What is worth noting is that the principal components' pattern does not justify the perception of a relation between intangibles, market share and profitability and/or liquidity.

The picture that is achieved by the above exercise is that, in the textile sector in Northern Italy, the firm's profile can be obtained by a random composition of three main components. That is to say, a company could be liquid, not necessarily profitable and with high investments in intangibles, with a meaningful market share.

Table 3.11   Principal components.

Components	Eigenvalues	Explained variance on total variance(%)	Cumulated variance explained (%)
COMP1	2.762	39.458	39.458
COMP2	1.827	26.098	65.556
COMP3	1.098	15.689	81.245
COMP4	0.835	11.922	93.167
COMP5	0.226	3.226	96.393
COMP6	0.172	2.453	98.846
COMP7	0.081	1.154	100.000
Total	7.000	100.000	

Table 3.12   Correlation coefficients between original variables and components.

Ratios	COMP1	COMP2	COMP3	R^2(communalities)	Singularity*
ROE	0.367	**0.875**	0.053	0.902	0.098
ROI	0.486	**0.798**	−0.100	0.883	0.117
CR	**0.874**	−0.395	0.057	0.923	0.077
QR	**0.885**	−0.314	−0.044	0.883	0.117
MTCI	**−0.892**	0.149	0.196	0.856	0.144
SHARE (%)	−0.055	0.259	**−0.734**	0.609	0.391
R&S (%)	−0.215	−0.286	**−0.709**	0.631	0.369

*Share of variable's variance left unexplained by the considered components.

Another company could be profiled in a completely different combination of the three components.

Given the pattern of the three components, a generic new firm j, belonging to the same population of the sample used here (sector, region, and size, for instance), could be profiled using these three 'fundamental' characteristics.

How can we calculate the value of the three components, starting from the original variables? Table 3.13 shows the coefficients that link the original variables to the new ones.

Table 3.13   The link between variables and components.

Original variables	COMP1	COMP2	COMP3
ROE	0.133	0.479	0.048
ROI	0.176	0.437	−0.091
CR	0.316	−0.216	0.052
QR	0.320	−0.172	−0.040
MTCI	−0.323	0.082	0.178
SHARE (%)	−0.020	0.142	−0.668
R&S (%)	−0.078	−0.156	−0.646

The table can be seen as a common output of linear regression analysis. Given a new ith observation, the jth component's value $S_{\text{comp } j}$ is calculated by summing up the original variables x_i multiplied by the coefficients, as shown below:

$$S_{comp1} = Roe_i \times 0,133 + Roi_i \times 0,176 + CR_i \times 0,316 + QR_i \times 0,320 - MTCI_i$$

$$\times (0,323) - SHARE_i \times (0,020) - R\&S_i \times (0,078)$$

This value is expressed in the same scale of the original variables, that is, it is not standardized. All the components are in the same scale, so they are comparable with one another in terms of mean (higher, lower) and variance (high/low relative variability). Very often, this is a desirable feature for the model builder. Principal components maintain the fundamental information on the level and variance of the original data. Therefore, principal components are suitable to be used as independent variables to estimate models, as all other variables used in LDA, logistic regression and/or cluster analysis. In this perspective, principal component analysis could be employed in model building as a way to pre-filter original variables, reducing their number, avoiding the noise of idiosyncratic information.

Now, consider 'factor analysis', which is similar to principal component; it is applied to describe observed variables in terms of fewer (unobserved) variables, known as 'factors'. The observed variables are modeled as linear combinations of the factors.

Why do we need factors? Unless the latent variable of the original q variables dataset is singular, the principal component analysis may not be efficient. In this case, factor analysis may be useful.

Assume that there are three 'true' latent variables. The principal component analysis attempts to extract the most common first component. This attempt may not be completed in an efficient way, because we know that there are three latent variables and each one will be biased by the effect of the other two. In the end, we will have an overvaluation of the first component contribution; in addition, its meaning will not be clear, because of the partial overlapping with the other two latent components. We can say that, when the likely number of latent variables is more than one, we will have problems in effectively finding the principal component profiles associated to the 'true' underlying fundamentals. So, the main problem of principal components analysis is to understand what the meaning of the new variables is, and to use them as more efficient combination of the original variables.

This problem can be overcome using the so called 'factor analysis', that is, in effect, often employed as the second stage of principal component analysis. The role of this statistical method is to:

- define the minimum statistical dimensions needed to efficiently summarize and describe the original dataset, free of information redundancies, duplications, overlapping, and inefficiencies;

- make a transformation of the original dataset, to give the better statistical meaning to the new latent variables, adopting an appropriate optimization algorithm to maximize the correlation with some variables and minimizing the correlation with others.

In this way, we are able to extract the best information from our original measures, understand them and reach a clear picture of what is hidden in our dataset and what is behind the borrowers' profiles that we directly observe in raw data.

Thurstone (1947), an American pioneer in the fields of psychometrics and psychophysics, described the set of criteria needed to define 'good' factor identification for the first time. In a correlation matrix showing coefficients between factors (in columns) and original variables (in rows), the required criteria are:

(a) each row ought to have at least one zero;

(b) each column ought to have at least one zero;

(c) considering the columns pair by pair, as many coefficients as possible have to be near zero in one variable and near one in the other; there should be a low number of variables with value near one;

(d) if factors are more than two, in many pairs of columns some variables have to be near zero in both columns.

In reality, these sought-after profiles are difficult to reach. To better target a factors' structure with these features, a further elaboration is needed; we can apply a method called 'factor rotation', a denomination derived from its geometrical interpretation. Actually, the operation could be thought of as a movement of the variable in the q-dimensions hyperspace to better fit some variables and to get rid of others, subject to the condition to have orthogonal factors one to the other. This process is a sort of factors adaptation in the space, aimed at better arranging the fit with the original variables and achieving more recognizable final factors.

To do this, factors have to be isomorphic, that is, standardized numbers, in order to be comparable and easily transformable. So, the first step is to standardize the principal components. Then, factor loadings (i.e., the value of the new variables) should be expressed as standardized figures (mean equal to zero and standard deviation equal to one). Factor loadings are comparable to one another but are not comparable (for range and size) with the original variables (on the contrary it is possible for principal components).

Furthermore, the factors depend on the criteria adopted to conduct the so-called 'rotation'. There are many criteria available. Among the different solutions available, there is the so called 'varimax method'[2]. This rotation method targets either large or small loadings of any particular variable for each factor. The method is based on an orthogonal movement of the factor axes, in order to maximize the variance of the squared loadings of a factor (column) on all of the variables (rows)

---

[2] Varimax rotation was introduced by Kaiser (1958) The alternative called 'normal-varimax' can also be considered. The difference is the use of a rotation weighted on the factor eigenvalues (Loehlin, 2003; Basilevsky, 1994). For a wider discussion on the Kaiser Criterion see Golder and Yeomans (1982).

in a factor matrix. The obtained effect is to differentiate the original variables by extracted factors. A varimax solution yields results which make it as easy as possible to identify each variable with a single factor. In practice, the result is reached by iteratively rotating factors in pairs; at the end of the iterative process, when the last round does not add any new benefit, the final solution is achieved. The Credit Risk Tracker model, developed by the Standards & Poor's rating agency for unlisted European and Western SME companies, uses this application[3].

Another example is an internal survey, conducted at Istituto Bancario Sanpaolo Group on 50 830 financial reports, extracted from a sample of more than 10 000 firms, collected between 1989 and 1992. Twenty-one ratios were calculated; they were the same used at that time by the bank to fill in credit approval forms; two dummy variables were added to take the type of business incorporation and the financial year into consideration. The survey objective was a preliminary data cleansing trying to identify clear, dominant profiles in the dataset, and separating 'outlier' units from the largely homogeneous population. The elaboration was based on a two-stage approach, the first consisting of factor analysis application, and the second using factors profiles to work out clusters of homogenous units.

Starting from 21 variables, 18 were the components with an eigenvalue of more than one, accounting for 99% of total variance; the first five, on which we will concentrate our analysis, accounted for 94%. Then, these 18 components were standardized and rotated. The explanation power was split more homogeneously through the various factors. The first five were confirmed as the most common and were able to summarize 42% of total variance in a well and identifiable way; the 'Cattell scree test' (that plots the factors on the X axis and the corresponding eigenvalues on the Y-axis in descending order) revealed a well established elbow after the first five factors and the others. The remaining 13 factors were rather a better specification of individual original attributes than factors which were able to summarize common latent variables. These applications were very useful, helping to apply at best cluster analysis that followed, conducted on borrowers' profiles based on common features and behaviors. Table 3.14 reports original variables, means, and factors structures, that is, the correlation coefficients between original variables and factors.

Coming to the economic meanings of the results of the analysis, it can be noted that the first six variables derive from classical ratios decomposition. Financial profitability, leverage, and turnover are correlated to three different, orthogonal factors. As a result, they are three different and statistically independent features in describing a firm's financial structure. This is an expected result from the firm's financial theory; for instance, from Modigliani–Miller assertions that separated operations from financial management. Moreover, from this factor analysis, assets turnover is split into two independent effects, that of fixed assets turnover on one side and that of working capital turnover on the other. This interpretation is very interesting. Very similar conclusions emerge from the PIMS econometric analysis, where capital intensity is proven to highly influence strategic choices and competitive positioning among incumbents and potential competitors, crucially

---

[3] Cangemi, De Servigny, and Friedman, 2003; De Servigny et al., 2004.

Table 3.14   Correlation among factors and variables in a sample of 50 830 financial reports (1989–1992).

Ratios	Means	Fact1	Fact2	Fact3	Fact4	Fact5
		Correlation coefficients (%)				
RoE	6.25%	43.9			−11.4	
RoI	7.59%	87.3		−20.3		
Total Leverage	5.91x		96.8		16.7	
Shareholders' Profit on industrial margin	−0.28%	37.2				
RoS	5.83%	94.6				
Total Assets Turnover	1.33×			−55.3		33.5
Gross Fixed Assets turnover	4.37×		21.3			89.9
Working Capital turnover	1.89×			−64.3		
Inventories turnover	14.96×			−12.5		
Receivables (in days)	111.47			96.8		
Payables (in days)	186.34		11.8	28.5		
Financial Leverage	4.96×		97.0		16.2	
Fixed assets coverage	1.60×		−16.9		−20.5	22.8
Depreciation level	54.26%				−18.5	
Sh/t Financial Gross Debt turnover	1.39×		−28.1	10.6	−48.7	
Sh/t net debt turnover	0.97×		−24.9	18.2	−44.7	
Sh/t debt on gross working capital	25.36%		12.2		97.0	
Sales (ITL.000.000) per employee	278.21			−14.8	22.2	
Added value per employee (ITL.000.000)	72.96	27.3				
Wages and Salaries per employee (ITL.000.000)	41.92					
Gross Fixed assets per employee (ITL.000.000)	115.05		−10.6			−18.9
Interest payments coverage	2.92×	49.6	−15.2		−26.2	
0 = partnership. 1 = stock company	0.35		11.7			10.2
Year end (1989, 1990, 1991, 1992)	1990.50					
% of variance explained by each factor		10.6	10.0	8.7	7.4	5.1
% cumulated variance explained		10.6	20.7	29.4	36.8	41.9

impacting on medium term profits and financial returns. The last factor regards the composition of firm's financial sources, partially influenced by the firm's competitive power in the customer/supply chain, with repercussions on leverage and liabilities arrangement.

Eventually, a final issue regards the economic cycle. The financial years from 1989 to 1992 were dramatically different. In particular, 1989 was one of the best years since the Second World War; 1992 was one of the worst for Italy, culminating with a dramatic devaluation of the currency, extraordinary policy measures and, consequently, the highest default rate in the industrial sectors recorded till now. We can note that the effect of the financial year is negligible, stating that the economic cycle is not as relevant as it is often assumed in determining the structural firms' profiles.

The cluster analysis that followed extracted 75% of borrowers with high statistical homogeneity and, based on them, a powerful discriminant function was estimated. The remaining 25% of borrowers showed high idiosyncratic behaviors, because they were start-ups, companies in liquidation, demergers or recent mergers; or simply data loading mistakes, or cases with too many missing values. By segregating these units, a high improvement in model building was achieved, avoiding statistical 'white noise' that could give unreliability to estimates.

The final part of this section is devoted to the so-called 'canonical correlation' method, introduced by Hotelling in the 1940s. This is a statistical technique used to work out the correspondence between a set of dependent variables and another set of independent variables. Actually, if we have two sets of variables, one dependent (Y_i) and another to explain the previous one (independent variables, X_i), then canonical correlation analysis enables us to find linear combinations of the Y_i and the X_i which have a maximum correlation with each other. Canonical correlation analysis is a sort of factor analysis in which the factors are extracted out of the X_i set, subject to the maximum correlation with the factors extracted out of the Y_i set. In this way we are able to work out:

- how many factors (i.e,. fundamental or 'basic' information) are embedded in the Y_i set,

- the corresponding factors out of the X_i set that are maximally correlated with factors extracted from Y_i set.

Y and X factors are orthogonal to one another, guaranteeing that we analyze actual (or latent) dimensions of phenomena underlying the original dataset.

In theory, canonical correlation can be a very powerful method. The only problem lies in the fact that, at the end of the analysis, we cannot rigorously calculate factors' scores, and, also, we cannot measure the borrowers' profile in new dependent and independent factors, but instead we can only generate proxies.

A canonical correlation is typically used to explore what is common amongst two sets of variables. For example, it may be interesting to explore what is explaining the default rate and the change of the default rate on different time horizons. By considering how the default rate factors are related to the financial ratios factors, we can gain an insight into what dimensions were common between the tests and how much variance was shared. This approach is very useful before starting to build a model based on two sets of variables; for example, a set of performance measures and a set of explanatory variables, or a set of outputs and a set of inputs. Constraints could also be imposed to ensure that this approach reflects the theoretical requirements.

Recently, on a database of internal ratings, in SanPaoloIMI a canonical correlation analysis has been developed. It aims at explaining actual ratings and changes (Y set) by financial ratios and qualitative attributes (X set). The results were interesting: 80% of the default probability (both in terms of level and changes) was explained by the first factor, based on high coefficients on default probabilities; 20% was explained by the second factor, focused only on changes in default probability. This second factor was highly correlated with a factor extracted from the X set, centered on industrial and financial profitability. The interpretation looks unambiguous: part of the future default probability change depends on the initial situation; the main force to modify this change lies in changes in profitability. A decline in operational profits is also seen as the main driver for the fall in credit quality and vice versa.

Methods like cluster analysis, principal component, factor analysis, and canonical correlation are undoubtedly very attractive because their potential contribution in the cleansing dataset and refining the data interpretation and the model building approach. Considering clusters, factors or canonical correlation structures helps to better master the information available and identify the borrower' profile determinants. Starting from the early 1980s, these methods achieved a growing role in statistics, leading to the so called 'exploratory multidimensional statistical analyses'; this was a branch of statistics born in the 1950s as *explorative statistics* (Tukey, 1977). He introduced the distinction between *exploratory data analysis* and *confirmatory data analysis*, and stated that the statistical analysis often gives too much importance to the latter, undervaluing the former. Subsequently, this discipline assumed a very relevant function in many fields, such as finance, health care, marketing, and complex systems' analysis (i.e., discovering the properties of complex structures, composed by interconnected parts that, as a whole, exhibit behaviors not obvious from the properties of the individual parts).

These methods are different in role and scope from discriminant or regression analyses. These last two methods are directly linked with decision theory (which aims at identifying values, uncertainties and other issues relevant for rational decision making). As a result of their properties, discriminant and regression analyses permit inferring properties of the 'universe' starting from samples. Techniques of variance reduction and association do not share these properties; they are not methods of optimal statistical decision. Their role is to arrange, order, and compact the available information, to reach better interpretations of information, as well as to avoid biases and inefficiencies in model building and testing. When principal components are used as a pre-processor to a model, their validity, stability and structure has to be tested over time in order to assess if solutions are still valid or not. In our experience, the life cycle of a principal component solution is around 18–24 months; following the end of this period, important adjustments would be needed.

Conversely, these methods are very suitable for numerical applications, and neural networks in particular, as we will subsequently examine.

3.3.8   Cash flow simulations

The firm's future cash flow simulation ideally stays in the middle between reduced form models and structural models. It is based on forecasting a firm's pro-forma financial reports and studying future performances' volatility; by having a default definition, for instance, we can see how many times, out of a set of iterative simulations, the default barrier will be crossed. The number of future scenarios in which default occurs, compared to the number of total scenarios simulated, can be assumed as a measure of default probability.

Models are based partly on statistics and partly on numeric simulations; the default definition could be exogenously or endogenously given, due to a model's aims and design. So, as previously mentioned, structural approaches (characterized by a well defined path to default, endogenously generated by the model) and reduced form approaches (characterized by exogenous assumptions on crucial variables, as market volatility, management behaviors, cost, and financial control and so forth) are mixed together in different model architectures and solutions.

It is very easy to understand the purposes of the method and its potential as a universal application. Nevertheless, there are a considerable number of critical points. The first is model risk. Each model is a simplification of reality; therefore, the cash flow generator module cannot be the best accurate description of possible future scenarios. But, the cash flow generator is crucial to count expected defaults. Imperfections or inaccuracies in its specification are vital in determining default probability. Hence, it is evident that we are merely transferring one problem (the direct determination of default probability through a statistical model) to another (the cash flow generator that produces the number of potential default circumstances). Moreover, future events have to be weighed by their occurrence in order to rigorously calculate default probabilities. In addition, there is the problem of defining what default is for the model. We do not know if and when a default is actually filed in real circumstances. Hence, we have to assume hypotheses about the default threshold. This threshold has to be:

- not too early, otherwise we will have many potential defaults, concluding that the transaction is very risky (but associated LGD will be low),

- not too late, otherwise we will have low default probability (showing a low risk transaction) but we could miss some pre-default or soft-default circumstances (LGD will be predicted as severe).

Finally, the analysis costs have to be taken into consideration. A cash flow simulation model is very often company specific or, at least, industry specific; it has to be calibrated with particular circumstances and supervised by the firm's management and a competent analyst. The model risk is amplified by the cost to build it, to verify and maintain its effectiveness over time. If we try to avoid (even partially) these costs, this could reduce the model efficiency and accuracy.

Despite these problems, we have no real alternatives to using a firm's simulation model in specific conditions, such as when we have to analyze a start-up, and we have no means to observe historical data. Let's also think about special purpose entities, project companies, or companies that have merged recently, LBOs or other situations in which we have to assess plans but not facts. Moreover, in these transactions covenants and 'negative pledges' are very often contractually signed to control specific risky events, putting lenders in a better position to promptly act in deteriorating circumstances. These contractual clauses have to be modeled and contingently assessed to verify both when they are triggered and what their effectiveness is. The primary cash flow source for debt repayment stays in operational profits and, in case of difficulties, the only other source of funds is company assets; usually, these are no-recourse transactions and any guarantee is offered by third parties. These deals have to be evaluated only against future plans, with no past history backing-up lenders. Individual analysis is needed, and it is necessarily expensive. Therefore, these are often 'big ticket' transactions, to spread fixed costs on large amounts. Rating (and therefore default probability) is assigned using cash flow simulation models.

These models are often based on codified steps, producing inter-temporal specifications of future pro-forma financial reports, taking into consideration scenarios regarding:

- how much cash flows (a) will be generated by operations, (b) will be used for financial obligations and other investments, and (c) what are their determinants (demand, costs, technology, and other crucial hypotheses)

- complete future pro-forma specifications (at least in the most advanced models), useful for also supporting more traditional analysis by ratios as well as for setting covenants and controls on specific balance sheet items.

To reach the probability of default, we can use either a scenario approach or a numerical simulation model. In the first case, we can apply probability to different (discrete) pre-defined scenarios. Rating will be determined through a weighted mathematical expectation on future outcomes; having a spectrum of future outcomes, defined by an associated probability of occurrence, we could also select a confidence level (e.g., 68% or 95%) to cautiously set our expectations. In the second case, we can use a large number of model iterations, which describe different scenarios: default and no-default (and also more diversified situations such as near-to-default, stressed and so forth) are determined and then the relative frequency of different stages is computed.

To give an example, Figure 3.7 depicts the architecture of a proprietary model developed for financial project applications, called SIMFLUX. The default criterion is defined in a Merton style approach; default occurs when the assets value falls below the 'debt barrier'. The model also proposes the market value of debt, showing all the intermediate stages (sharp reduction in debt value) in which repayment is challenging but still achievable (near-to-default stages). These situations are very important in financial projects because, very often, waivers are forced if things turn out badly, in order to minimize non-payment and default filing that would generate credit losses.

The model works as follows:

- the impact on project outcomes is measured, based on industrial valuations, sector perspectives and analysis, key success factors and so forth. Sensitivity to crucial macro-economic variables is then estimated. Correlation among the

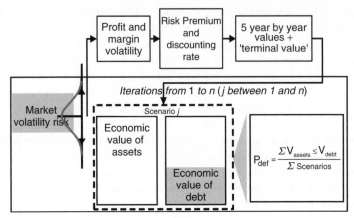

Figure 3.7   SIMFLUX cash flow simulation model architecture. Source: *internally developed.*

macro-economic risk factors is ascertained in order to find joint probabilities of potential future outcomes (scenario engine);

- given macroeconomic joint probabilities, random scenarios are simulated to generate revenues' volatility and its probability density function;

- applying the operational leverage, margin volatility is estimated as well. Then, the discount rate is calculated, with regards to the market risk premium and business volatility;

- applying the discount rate to cash flows, the firm's value is produced (from time to time for the first five years plus 'terminal value' beyond this horizon, using an asymptotic 'fading factor' for margins growth);

- Monte Carlo random simulations are then run, to generate the final expected spectrum of assets and debt value;

- then, default frequencies are counted, that is, the number of occurrences in which assets values are less than debt values. Consequently, a probability of default is determined;

- debt market values are also utilized by the model, plotted on a graph, to directly assess when there is a significant reduction in debt value, indicating difficulties and potential 'near-to-default' situations.

3.3.9   A synthetic vision of quantitative-based statistical models

Table 3.15 shows a summary valuation on quantitative statistical-based methods to ratings assignment, mapped against the three desirable features previously described.

Table 3.15 Overview of quantitative based statistical ratings.

Criteria	Structural approach	Reduced form approaches			Cash flow simulations
	Option approach applied to stock listed companies	Discriminant analysis	Logistic regression	Unsupervised techniques*	
Measurability and verifiability	●	●	●	◕	◕
Objectivity and homogeneity	●	◕	◑	◐	◕
Specificity	◕	●	●	◔	◐

*Cluster analysis, principal components, factor analysis, canonical correlation.

Structural approaches are typically applied to listed companies due to the input data which is required. Variance reduction techniques are generally not seen as an alternative to regression or discriminant functions but rather as a complement of them: only cluster analysis can be considered as an alternative when top down approaches are preferred; this is the case when a limited range of data representing borrowers' characteristics is available. Cash flow analysis is used to rate companies whose track records are meaningless or non-existent. Discriminant and regression analyses are the principal techniques for bottom up statistical based rating models.

3.4   Heuristic and numerical approaches

In recent years, other techniques besides statistical analyses have been applied to default prediction; they are mostly driven by the application of artificial intelligence methods. These methods completely change the approach to traditional problem solving methods based on decision theory. There are two main approaches used in credit risk management:

(a) 'Heuristic methods', which essentially mimic human decision making procedures, applying properly calibrated rules in order to achieve solutions in complex environments. New knowledge is generated on a trial by error basis, rather than by statistical modeling; efficiency and speed of calculation are critical. These methods are opposed to algorithms-based approaches and are often known as 'expert systems' based on artificial intelligence techniques. The aim is to reproduce high frequency standardized decisions at the best level of quality and adopting low cost processes. Feedbacks are used to continuously train the heuristic system, which learns from errors and successes.

(b) 'Numerical methods', whose objective is to reach optimal solutions adopting 'trained' algorithms to take decisions in highly complex environments characterized by inefficient, redundant, and fuzzy information. One example of these approaches is 'Neural networks': these are able to continuously auto-update themselves in order to adjust to environmental modifications. Efficiency criteria are externally given or endogenously defined by the system itself.

3.4.1   Expert systems

Essentially, expert systems are software solutions that attempt to provide an answer to problems where human experts would need to be consulted. Expert systems are traditional applications of artificial intelligence. A wide variety of methods can be used to simulate the performance of an expert. Elements common to most or all expert systems are:

- the creation of a knowledge base (in other words, they are *knowledge-based systems*),

- the process of gathering knowledge and codifying it according to some frameworks (this is called knowledge engineering).

Hence, expert systems' typical components are:

1. the knowledge base,

2. the working memory,

3. the inferential engine,

4. the user's interface and communication.

The knowledge base is also known as 'long term memory' because it is the set of rules used for decisions making processes. Its structure is very similar to a database containing facts, measures, and rules, which are useful to tackle a new decision using previous (successful) experiences. The typical formalization is based on 'production rules', that is, 'if/then' hierarchical items, often integrated by probabilities p and utility u. These rules create a decision making environment that emulates human problem solving approaches. The speed of computers allows the application of these decision processes with high frequency in various contexts and circumstances, in a reliable and cost effective way.

The production of these rules is developed by specialists known as 'knowledge engineers'. Their role is to formalize the decision process, encapsulating the decision making logics and information needs taken from practitioners who are experts in the field, and finally combining different rules in layers of inter-depending steps or in decisional trees.

The 'working memory' (also known as short term memory) contains information on the problem to be solved and is, therefore, the virtual space in which rules are combined and where final solutions are produced. In recent years, information systems are no longer a constraint to the application of these techniques; computers' data storage capacity has increased to a point where it is possible to run certain types of simple expert systems even on personal computers.

The inferential engine, at the same time, is the heart and the nervous network of an expert system. An understanding of the 'inference rules' is important to comprehend how expert systems work and what they are useful for. Rules give expert systems the ability to find solutions to diagnostic and prescriptive problems. An expert system's rule-base is made up of many inference rules. They are entered into the knowledge base as separate rules and the inference engine uses them together to draw conclusions. As each rule is a unit, rules may be deleted or added without affecting other rules. One advantage of inference rules over traditional models is that inference rules more closely resemble human behavior. Thus, when a conclusion is drawn, it is possible to understand how this conclusion was reached. Furthermore, because the expert system uses information in a similar manner to experts, it may be easier to find out the needed information from banks' files. Rules can also incorporate probability of events and the gain/cost of them (utility).

The inferential engine may use two different approaches, backward chaining and forward chaining respectively:

- 'Forward chaining' starts with available data. Inference rules are used until a desired goal is reached. An inference engine searches through the inference rules until it finds a solution that is pre-defined as correct; the path, once recognized as successful, is then applied to data. Because available data determine which inference rules are used, this method is also known as data driven.

- 'Backward chaining' starts with a list of goals. Then, working backwards, the system tries to find the path which allows it to achieve any of these goals. An inferential engine using backward chaining would search through the rules until it finds the rule which best matches a desired goal. Because the list of goals determines which rules are selected and used, this method is also known as goal driven.

Using chaining methods, expert system can also explore new paths in order to optimize target solutions over time.

Expert systems may also include fuzzy logic applications. Fuzzy logic has been applied to many fields, from control theory to artificial intelligence. In default risk analysis, many rules are simply 'rule-of-thumb' that have been derived from experts' own feelings; often, thresholds are set for ratios but, because of the complexity of real world, they can result to be both sharp and severe in many circumstances. Fuzzy logic is derived from 'fuzzy set theory', which is able to deal with approximate rather than precise reasoning. Fuzzy logic variables are not constrained to the two classic extremes of black and white logic (zero and one), but rather they may assume any value between the extremes. When there are several rules, the set thresholds can hide incoherencies or contradictions because of overlapping areas of uncertainty and logical mutual exclusions. Instead, adopting a more flexible approach, many clues can be integrated, reaching a solution that converges to a sounder final judgment. For example:

- if interest coverage ratio (EBIT divided by interest paid) is less than 1.5, the company is considered as risky,

- if ROS (EBIT divided by revenues) is more than 20%, the company is considered to be safe.

The two rules can be combined together. Only when both are valid, that is to say ROS is lower (higher) than 20% and interest coverage is lower (higher) than 1.5, we can reach a dichotomous risky/safe solution. In all other cases, we are uncertain.

When using the fuzzy logic approach, the 'low interest coverage rule' may assume different levels depending on ROS. So, when a highly profitable company is considered, less safety in interest coverage can be accepted (for instance, 1.2). Therefore, fuzzy logic widens the spectrum of rules that expert systems can use, allowing them to approximate human decisional processes even better.

Expert systems were created to substitute human-based processes by applying mechanical and automatic tools. When knowledge is well consolidated and stabilized, characterized by frequent (complex and recursive) calculations and associated with well established decision rules, then expert systems are doing their best in exploring all possible solutions (may be millions) and in finding out the best one. Over time, their knowledge base has extended to also include ordinal and qualitative information as well as combinations of statistical models, numerical methods, complex algorithms, and logic/hierarchical patterns of many interconnected submodels. Nowadays, expert systems are more than just a way to solve problems or to model some real world phenomena; they are software that connects many subprocesses and procedures, each optimized in relation to its goals using different rules. Occasionally, expert systems are also used when there are completely new conditions unknown to the human experience (new products, new markets, new procedures, and so forth). In these cases, as there is no expertise, we need to explore what can be achieved by applying rules derived from other contexts and by following a heuristic approach.

In the credit risk management field, an expert system based on fuzzy logic used by the German Bundesbank since 1999 (Blochwitz and Eigermann, 2000) is worth noting. It was used in combination with discriminant analysis to investigate companies that were classified by the discriminant model in the so-called 'gray area' (uncertain attribution to defaulting/performing classes). The application of the expert system raised the accuracy from 18.7% of misclassified cases by discriminant function to an error rate of only 16% for the overall model (Figure 3.8).

In the early 1990s, SanPaoloIMI also built an expert system for credit valuation purposes, based on approximately 600 formal rules, 50 financial ratios, and

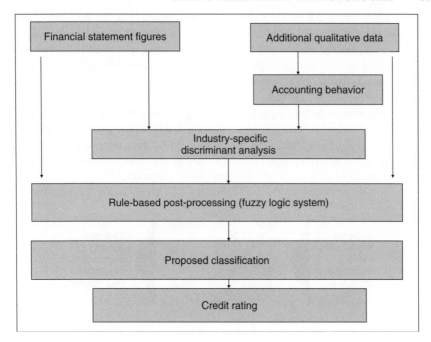

Figure 3.8   Expert system in Bundesbank's credit quality valuation model.

three areas of analysis. According to a blind test, the system proved to be very efficient and guaranteed homogeneity and a good quality of results. Students who were studying economics (skilled in credit and finance) and students studying engineering (completely unskilled in credit and finance) were asked to separately apply the expert system to about 100 credit dossiers without any interaction. The final result was remarkable. The accuracy in data loading and in output produced, the technical comments on borrowers' conditions, and the time employed, were the same for both groups of skilled and unskilled people. In addition, the borrowers' defaulting or performing classifications were identical because they were strictly depending on the model itself. These are, at the same time, very clearly, the limits and opportunities of expert systems.

Decision Support Systems (DSSs) are a subset of expert systems. These are models applied to some phases of the human decision process, which mostly require cumbersome and complex calculations. DSSs have had a certain success in the past, also as stand-alone solutions, when computer power was quickly increasing. Today, many DSSs applications are part of complex procedures, supporting credit approval processes and commercial relationships.

3.4.2 Neural networks

Artificial neural networks originate from biological studies and aim to simulate the behavior of the human brain, or at least a part of the biological nervous system

(Arbib, 1995; Steeb, 2008). They comprise interconnecting artificial neurons, which are software programs intended to mimic the properties of biological neurons.

Artificial neurons are hierarchical 'nodes' (or steps) connected in a network by mathematical models that are able to exploit connections by operating a mathematical transformation of information at each node, often adopting a fuzzy logic approach. In Figure 3.9, the network is reduced to its core, that is, three layers. The first is delegated to handle inputs, pre-filtering information, stimuli, and signals. The second (hidden) is devoted to computing relationships and aggregations; in more complex neural networks this component could have many layers. The third is designated to generate outputs and to manage the users' interface, delivering results to the following processes (human or still automatic).

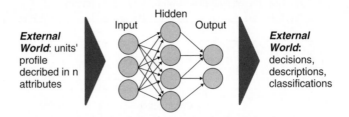

Figure 3.9   Frame of a neural network.

Thanks to the (often complex) network of many nodes, the system is able to fit into many different cases and can also describe nonlinear relationships in a very flexible way. After the 'initial training', the system can also improve its adaptability, learning from its successes and failures over time.

To better clarify these concepts, compare neural networks with traditional statistical analyses such as regression analysis. In a regression model, data are fitted through a specified relationship, usually linear. The model is made up by one or more equations, in which each of the inputs x_j is multiplied by a weight w_j. Consequently, the sum of all such products and of a constant α gives an estimate of the output. This formulation is stable over time and can only be changed by an external decision of the model builder.

In neural networks, the input data x_j is again multiplied by weights (also defined as the 'intensity' or 'potential' of the specific neuron), but the sum of all these products is influenced by:

- the argument of a flexible mathematical function (e.g., hyperbolic tangent or logistic function),

- the specific calculation path that involves some nodes, while ignoring others.

The network calculates the signals gathered and applies a defined weight to inputs at each node. If a specific threshold is overcome, the neuron is 'active' and generates an input to other nodes, otherwise it is ignored. Neurons can interact with strong or weak connections. These connections are based on weights and on paths that inputs have to go through before arriving to the specific neuron. Some paths are privileged; however neurons never sleep: inputs always go through the entire network. If new information arrives, the network can search for new solutions testing other paths, and thus activating certain neurons while switching off others. Some paths could automatically substitute others because of the change in input intensity or in inputs profile. Often, we are not able to perceive these new paths because the network is always 'awake', immediately catching news and continuously changing neural distribution of stimuli and reactions. Therefore, the output y is a nonlinear function of x_j. So, the neural network method is able to capture nonlinear relationships.

Neural networks could have thousands of nodes and, therefore, tens of thousands of potential connections. This gives great flexibility to the whole process to tackle very complex, interdependent, no linear, and recursive problems.

The most commonly used structure is the 'hierarchically dependent neural network'. Each neuron is connected with previous nodes and delivers inputs to the following node, with no return and feedbacks, in a continuous and ordered flow. The final result is, therefore, the nonlinear weighted sum of inputs and defined as:

$$f(x) = k \left(\sum_i w_i g_i(x) \right)$$

where k is a pre-defined function; for instance, the logistic one. The ith neuron gathers stimuli from j previous neurons. Based on weights, the 'potential', v_i is calculated in a way depicted in Figure 3.10.

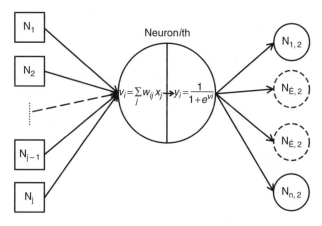

Figure 3.10   Communications among artificial neurons.

The potential is not comparable among neurons. A conversion is needed in order to compare them; the logistic conversion indicated in Figure 3.10 sets the output value between 0 and 1. When there is only one hidden layer, the neural network behaves like a traditional statistical logistic function. Unless very complex problems are being dealt with, one or two layers are enough to solve most issues, as also proven by mathematical demonstrations.

The final result of the neural network depends more on training than on the complexity of the structure.

Now, let's come to the most interesting feature of a neural network, the ability to continuously learn by experience (neural networks are part of 'learning machines'). There are different learning methods and many algorithms for training neural networks. Most of them can be viewed as a straightforward application of the optimization theory and statistical estimation.

In the field of credit risk, the most applied method is 'supervised learning', in which the training set is given and the neural network learns how to reach a successful result by finding the nodes' structure and the optimal path to reach the best final result. This also implies that a cost function is set in order to define the utility of each outcome. In the case of default risk model building, the training set is formed by borrowers' characteristics and the cost function reflects misclassification costs. A back-propagation learning engine may be launched to train the neural network. After much iteration, a solution that minimizes the classification errors is reached by changing the weights and connections at different nodes. If the training process is successful, the neural network learns the connections among inputs and outputs, and can be used to make previsions for new borrowers that were not present in the training set. Accuracy tests are to be performed to gauge if the network is really able to solve problems in out-of-sample populations, with an adequate level of generality.

The new generations of neural networks are more and more entwined with statistical models and numerical methods. Neural networks are (apparently) easy to use and generate a workable solution fairly quickly. This could set a mental trap: complex problems remain complex even if a machine generates adequate results; competences in statistics and a good control of the information set are unavoidable.

The main limit of neural networks is that we have to accept results from a 'black box'. We cannot examine step by step how results are obtained. Results have to be accepted as they are. In other words, we are not able to explain why we arrive at a given result. The only possibility is to prepare various sets of data, well characterized with some distinguishing profiles, then submit them to the neural network to reach results. In this case, by having outputs corresponding to homogenous inputs and using the system theory, we can deduce which the crucial variables are and their relative weights.

Much like any other model, neural networks are very sensitive to input quality. So, training datasets have to be carefully selected in order to avoid training the model to learn from outliers instead of normal cases. Another limit is related to the use of qualitative variables. Neural network are more suited to work with continuous quantitative variables. If we use qualitative variables, it is advisable to avoid dichotomous categorical variables, preferring multiple ranked modalities.

There are no robust scientific ways to assess if a neural network is optimally estimated (after the training process). The judgment on the quality of the final neural network structure is mainly a matter of experience, largely depending on the choices of knowledge engineers made during model building stages.

The major danger in estimating neural networks is the risk of 'over-fitting'. This is a sort of network over-specialization in interpreting the training sample; the network becomes completely dependent on the specific training set. A network that over-fits a sample is incapable of producing satisfactory results when applied to other borrowers, sectors, geographical areas or economic cycle stages. Unfortunately, there are no tests or techniques to gauge if the solution is actually over-fitting or not. The only way out is to use practical solutions. On one hand, the neural network has to be applied to out-of-sample, out-of-time and out-of-universe datasets to verify if there are significant falls in statistical performances. On the other hand, the neural network has to be continuously challenged, very often re-launching the training process, changing the training set, and avoiding specialization in only one or few samples.

In reality, neural networks are mainly used where decisions are taken in a fuzzy environment, when data are rough, sometimes partially missed, unreliable, or mistaken. Another elective realm of application lies where dominant approaches are not provided, because of the complexity, novelty or rapid changes in external conditions. Neural networks are, for instance, nowadays used in negotiation platforms. They react very quickly to changing market conditions. Their added value clearly stays in a prompt adaptation to structural changes.

3.4.3   Comparison of heuristic and numerical approaches

Expert systems offer advantages when human experts' experience is clear, known, and well dominated. This enables knowledge engineers to formalize rules and build effective systems. Expert systems are ideal for the following reasons:

- to give order and structure to real life procedures, which allow decision making processes to be replicated with high frequency and robust quality;

- to connect different steps of decision making processes to one another, linking statistical and inferential engines, procedures, classifications, and human involvement together, sometimes reaching the extreme of producing outputs in natural language.

From our perspective (rating assignment), expert systems have the distinct advantage of giving order, objectivity, and discipline to the rating process; they are desirable features in some circumstances but they are not decisive. In reality, rating assessment depends more on models that are implanted inside the expert system, models that are very often derived from other methods (statistical, numerical), with their own strength and weaknesses. As a result, expert systems organize knowledge and processes but they do not produce new knowledge because they are not models or inferential methods.

Numerical algorithms such as neural networks have completely different profiles. Some applications perform quite satisfactorily and count many real life applications. Their limits are completely different and are mainly attributable to the fact that they are not statistical models and do not produce a probability of default. The output is a classification, sometimes with a very low granularity (four classes, for instance, such as very good, pass, verify, reject). To extract a probability, we need to associate a measure of default frequency obtained from historical data to each class. Only in some advanced applications, does the use of models implanted in the inferential engine (i.e., a logistic function) generate a probability of default. This disadvantage, added to the 'black box nature' of the method, limits the diffusion of neural networks out of consumer credit or personal loan segments.

However, it is important to mention their potential application in early warning activities and in credit quality monitoring. These activities are applied to data generated by very different files (in terms of structure, use, and scope) that need to be monitored frequently (i.e., daily or even intraday for internal behavioral data). Many databases often derived from production processes frequently change their structure. Neural networks are very suitable in going through a massive quantity of data, changing rapidly when important discontinuities occur, and quickly developing new rules when a changing pattern of success/failure is detected.

Finally, it should be noted that some people classify neural networks among statistical methods and not among numerical methods, because some nodes can be based on statistical models (for instance, ONB and FMA, 2004).

Table 3.16 offers the usual final graphic summary.

Table 3.16   An overview of heuristic and numerical based ratings.

Criteria	Heuristic approach	Numerical approach
	Expert systems / Decision support system	Neural networks
Measurability and verifiability		
Objectivity and homogeneity		
Specificity		

3.5   Involving qualitative information

Statistical methods are well suited to manage quantitative data. However, useful information for assessing probability of default is not only quantitative. Other types of information are also highly relevant such as: sectors competitive forces characteristics, firms' competitive strengths and weaknesses, management quality, cohesion and stability of entrepreneurs and owners, managerial reputation, succession plans in case of managerial/entrepreneurial resignation or turnaround, strategic continuity, regulations on product quality, consumers' protection rules and risks, industrial cost structures, unionization, non-quantitative financial risk profiles (existence of contingent plans on liquidity and debt repayment, dependency from the financial group's support, strategy on financing growth, restructuring and repositioning) and so forth; these usually have a large role in judgment-based approaches to credit approval and can be classified in three large categories:

i. efficiency and effectiveness of internal processes (production, administration, marketing, post-marketing, and control);

ii. investment, technology, and innovation;

iii. human resource management, talent valorization, key resources retention, and motivation.

In more depth:

- domestic market, product/service range, firm's and entrepreneurial history and perspectives;

- main suppliers and customers, both in terms of quality and concentration;

- commercial network, marketing organization, presence in regions and countries, potential diversification;

- entrepreneurial and managerial quality, experience, competence;

- group organization, like group structure, objectives and nature of different group's entities, main interests, diversification in non-core activities, if any;

- investments in progress, their final foreseeable results in maintaining/re-launching competitive advantages, plans, and programs for future competitiveness;

- internal organization and functions, resources allocation, managerial power, internal composition among different branches (administration, production, marketing, R&D and so forth);

- past use of extraordinary measures, like government support, public wage integration, foreclosures, payments delays, credit losses and so forth;

- financial relationships (how many banks are involved, quality of relationships, transparency, fairness and correctness, and so on);

- use of innovative technologies in payment systems, integration with administration, accounting, and managerial information systems;

- quality of financial reports, accounting systems, auditors, span of information and transparency, internal controls, managerial support, internal reporting and so on.

Presently, new items have become of particular importance: environmental compliance and conformity, social responsibility, corporate governance, internal checks and balances, minorities protection, hidden liabilities like pension funds integration, stock options and so on.

A recent summing up of usual qualitative information conducted in the Sanpaolo Group in 2006 collected more than 250 questions used for credit approval processes, extracted from the available documents and derived from industrial and financial economy, theories of industrial competition and credit analysis practices. A summary is given in Table 3.17.

Qualitative variables are potentially numerous and, consequently, some ordering criterion is needed to avoid complex calculations and information overlapping. Moreover, forms to be filled in soon become very complex and difficult to be understood by analysts. A first recommendation is to only gather qualitative information that is not collectable in quantitative terms. For instance, growth and financial structure information can be extracted from balance sheets.

A second recommendation regards how to manage qualitative information in quantitative models. A preliminary distinction is needed between different categorical types of information:

- nominal information, such as regions of incorporation;

- binary information (yes/no, presence/absence of an attribute);

- ordinal classification, with some graduation (linear or nonlinear) in the levels (for instance, very low/low/medium/high/very high).

Binary indicators can be transformed in 0/1 '*dummy variables*'. Also, ordinal indicators can be transformed into numbers and weights can be assigned to different modalities (the choice of weights is, however, debatable).

When collecting data, it is preferable to structure the information in closed form if we want to use it in quantitative models. This means forcing loan officers to select some pre-defined answers.

Binary variables are difficult to managed in statistical models because of their non-normal distribution. Where possible, a multistage answer is preferable, instead of yes/no. Weights can be set using optimization techniques, like '*bootstrap*' or a preliminary test on different solutions to select the most suited one.

Nowadays, however, the major problem in using qualitative information lies in the lack of historical datasets. The credit dossier is often based on literary presentations without a structured compulsory basic scheme. Launching an extraordinary survey in order to collect missing information has generally proven to be:

- A very expensive solution. There are thousands of dossiers and it takes a long time for analysts to go through them. Final results may be inaccurate because they are generated under pressure.

- A questionable approach for non-performing dossiers. Loan and credit officers that are asked to give judgments on the situation as it was before the default are tempted to skip on weaknesses and to hide true motivations of their (ex post proved wrong) judgments.

Table 3.17   Example of qualitative items in credit analysis questionnaires.

- Corporate Structure

  - date of incorporation of the company(or of a significant merger and/or acquisition)
  - group members, intensity of relationship with the parent/subsidiary

- Information on the Company's business

  - markets in which the company operates and their position in the 'business life cycle' (introduction, growth, consolidation, decline)
  - positions with competitors and competitive strength
  - nature of competitive advantage (cost, differentiation/distinctiveness of products, quality/innovation/technology, dominant/defendable)
  - years the company operates in the actual core business
  - growth forecast
  - quality of the references in the marketplace

- Strategy

  - strategic plans
  - business plan
  - in case a business plan has been developed, the stage of strategy implementation
  - proportion of assets/investments not strategically linked to the company's business
  - extraordinary transactions (revaluations, mergers, divisions, transfers of business divisions, demerger of business) and their objective

- Quality of management

  - degree of involvement in the ownership and management of the company
  - the overall assessment of management's knowledge, experience, qualifications and competence (in relation to competitors)
  - if the company's future is tied to key figures
  - presence of a dominant entrepreneur/investor (or a coordinated and cohesive group of investors) that influence strategies and company's critical choices

- Other risks

  - risks related to commercial activity
  - geographical focus (the local/regional, domestic, within Europe, OECD and non-OECD/emerging markets)
  - level of business diversification (a single product/service, more products, services, markets)
  - liquidity of inventories
  - quality of client base

(continued)

Table 3.17   *(continued)*

- share of total revenues generated by the first three/five customers of the company
- exclusivity or prevalence with some company's suppliers
- legal and / or environmental risks
- reserves against professional risks, board members responsibilities, auditors (or equivalent insurance)

• Sustainability of financial position

- reimbursements within the next 12 months, 18 months, 3 years, and concentration of any significant debt maturities
- off-balance-sheet positions and motivations (coverage, management, speculation, other)
- sustainability of critical deadlines with internal/external sources and contingency plans
- liquidity risk, potential loss in receivables of one or more major customers, potential need to accelerate the payment of the most important suppliers)

• Quality of information provided by the company to the bank, timing in the documentation released and general quality of relationships

- availability of plausible financial projections
- information submitted on company's results and projections
- considerations released by auditors on the quality of budgetary information
- relationship vintage, past litigation, type of relation (privileged/strategic or tactical/opportunistic)
- managerial attention
- negative signals in the relationship history

There is no easy way to overcome these problems. A possible way is to prepare a two-stage process:

• The first stage is devoted to building a quantitative model accompanied by the launch of a systematic qualitative data collection on new dossiers. This qualitative information can immediately be used in overriding quantitative model results through a formal or informal procedure.

• The second stage is to build a new model including the new qualitative information gathered once the first stage has produced enough information (presumably after at least three years), trying to find the most meaningful data and possibly re-engineering the data collection form if needed.

Note that qualitative information change weight and meaningfulness over time. At the end of 1980s, for instance, one of the most discriminant variables was to operate or not on international markets. After the globalization, this feature is less important; instead, technology, marketing skills, brands, quality, and management competences have become crucial. Therefore, today, a well structured and reliable qualitative dataset is an important competitive hedge for banks, an important component to build powerful credit models, and a driver of banks' long term value creation.

4

Developing a statistical-based rating system

4.1 The process

In this chapter, the statistical-based component of a rating system (in short, SBRS), also referred to as a 'scoring model', is developed. The process of development is divided into many stages, steps, and substeps. Even if the different nature of data (e.g., continuous or categorical) affects the details of the process, basic individual steps for quantitative and qualitative data, as well as for their subtypes (balance sheet, internal behavioral, credit register, business sector, company's competitive positioning, etc.) have many features in common. In order to avoid repetition, the chapter is primarily based on quantitative data analysis and, in particular, on balance sheet and income statement data. A summary of stages, steps, and substeps of model development is illustrated in Table 4.1.

Setting the model's objectives is the first decision to take. Some aspects are clear cut and require an immediate decision because of their evident impact on all subsequent stages, steps, and substeps.

- Will the model be used for evaluating new customers or for reviewing existing customers' creditworthiness? In the former case, internal behavioral data are not available.

- Will the model be applied to individuals, financial or non-financial firms? The structure of accounting data is completely different in each case.

- Will the model be used for daily monitoring or rather for long term lending decisions? The time frame of the data to be considered is different from one another.

Developing, Validating and Using Internal Ratings: Methodologies and Case Studies Giacomo De Laurentis, Renato Maino and Luca Molteni © 2010 John Wiley & Sons, Ltd

Table 4.1   Stages, steps, and substeps of model development.

---

1 Setting model's objectives and nature of data to be collected
2 Setting the time-frame of data and generating the dataset
3 Preliminary analysis

    3.1 The dataset: an overview
    3.2 Duplicate cases analysis
    3.3 Missing values analysis
    3.4 Missing value treatment
    3.5 Other preliminary overviews

4 Defining an analysis sample

    4.1 Rationale for splitting the dataset into an analysis sample and a validation sample
    4.2 How to split the dataset into an analysis sample and a validation sample

5 Univariate and bivariate analyses

    5.1 Indicators' economic meanings, working hypotheses and structural monotonicity

        5.1.1 Scale variables
        5.1.2 Categorical variables

    5.2 Empirical assessment of working hypothesis
    5.3 Normality and homogeneity of variance
    5.4 Graphical analysis
    5.5 Discriminant power
    5.6 Empirical monotonicity
    5.7 Correlations
    5.8 Analysis of outliers
    5.9 Transformation of indicators

        5.9.1 Treatment of outliers and shapes of distributions
        5.9.2 Treatment of empirical non-monotonicity
        5.9.3 Other transformation: categorical variables

    5.10 Indicators short listing

6 Estimating a model and assessing its discriminatory power

    6.1 Steps and case study simplifications
    6.2 Linear discriminant analysis
    6.3 Logistic regression

7 From scores to ratings and from ratings to probabilities of default

---

• Is the model's aim to assign ratings or to solve the dichotomy accept/reject? The final structure of the model and the methods of performance evaluations would be altogether different.

• Is it better to develop different scoring functions for different types of data (e.g., balance sheet, internal behavioral, credit register, qualitative, etc.) and then combine these modules in a final model or, alternatively, would it be better to estimate the model using all sources of information at once? Model structure will be heavily impacted by this decision.

Some other aspects could be less clear cut in the outset of the analysis and could become clearer in the process. Is it necessary to develop specialized models for real estate companies? The model's performance on the specific segment may provide the best answer.

In conclusion, setting the model's objectives determines the collection of data and shapes the features of the model building process.

Generating the dataset is often the most time consuming and expensive stage of the process (fortunately we will avoid this stage in the case studies presented in this chapter). It is also vital for the integrity of the model that the empirical data satisfy some important requirements, such as representativeness of the segment to which the model will be applied, as well as quantity (to enable statistically significant statements) and quality (to avoid distortions due to unreliable data).

Preliminary analysis on the collected dataset has three objectives: (i) to have a first glance at the available data, (ii) to erase duplicate cases, and (iii) and to identify the nature of missing values and their quantitative relevance.

Purposes of univariate and bivariate analyses are manifold:

1. to identify creditworthiness characteristics which make sense in the business context;

2. to check that central tendency measures (such as media and median for the two subsamples of defaulted and non-defaulted observations) are in line with working hypotheses;

3. for the two said subsamples, to analyze if the entire distribution of each indicator's values are separated enough. Graphical analysis can be of help. It also shows the trend of distributions and the presence of outliers and extreme values;

4. to verify the frequency of outliers and extreme values, their origin and their impact;

5. to control the monotonic relationship between an indicator and default probability, both empirically and at a theoretical level;

6. to check for normality within the indicators' distributions and to check for homogeneity of variance in the two subsamples; these last two checks are essential because many multivariate methodologies require such characteristics for independent variables;

7. to quantify the discriminatory power of individual indicators;

8. to study the (linear and non-parametric) correlation among pairs of indicators, in order to leave out those which are highly correlated with other more powerful indicators.

Once these analyses have been undertaken, the opportunity to transform some indicators in order to solve problems such as monotonicity, normality, and/or reduction in the impact of outliers and extremes possibly emerges. If indicators are transformed, the previously listed analyses should be carried out again.

The results of univariate and bivariate analyses can be conveniently summarized into a synoptic table, allowing for a final decision to be made on which indicators to enter into the short list.

4.2   Setting the model's objectives and generating the dataset

4.2.1   Objectives and nature of data to be collected

The previous paragraph has examined the model's objectives and the nature of data to be collected. These are preliminary choices for any model development. As such, they constitute the first stage of the process. The following paragraph is devoted to an in-depth analysis of the time frame of data to attain in the collection process.

4.2.2   The time frame of data

A key aspect is to set a precise time frame by which information is collected (Figure 4.1). 'Time zero' is the theoretical point in time in which credit analysis is carried out, so all information available at that moment can be incorporated into the model, which will attempt to forecast whether the borrower will go into default during the observation period. Behavioral data produced by bank information systems are available on the following day. Credit register data are available according to the timing of data distribution set up by the credit register; in Italy, on average they are available after approximately 1.5 months. If yearly balance sheets, income statements, and flow of funds statements are approved after a few months from year-end, they become available for credit analyses that take place, on average, about one year after year-end. Quarterly results can minimize this time lag. A significant part of SBRS considers a default observation period of one year from time zero.

Considering a default observation period of more than one year would lead to building more forward looking models. The relevant explanatory variables in the models can change considerably. However, performance measures of discriminatory power and proper calibration worsen. So, a critical decision should be made. Due to the 2008 financial crisis, banks have very recently started to experiment

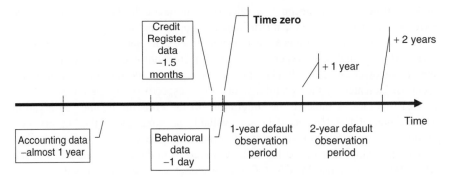

Figure 4.1   Time frame for collecting data.

with a longer time horizon for estimating models and assigning rating, eventually moving towards an explicit but largely ignored Basel II requirement: 'Although the time horizon used in PD estimation is one year (as described in paragraph 447), banks are expected to use a longer time horizon in assigning ratings' (Basel Committee, 2004, §414).

In any case, the time frame linking a given default observation period and corresponding data available at time zero must be respected. However, it is possible to collect observations of different borrowers to include within a dataset using different time-zeros. This time-stratification approach is welcomed as it decreases the dependence of data from a particular calendar year, consequently smoothing the relation between the dataset and a specific stage of the economic cycle.

4.3   Case study: dataset and preliminary analysis

4.3.1   The dataset: an overview

The case study aims at matching the theory of model building in the field of borrower ratings and the practice of managing data and statistical tools such as SPSS-PASW. Initially, basics of SPSS-PASW structure and routines are given. Therefore, the reader that already has these rudiments is invited to advance quickly, picking up only the elements that are critical for dataset structuring and model building.

The task is to build the statistical-based component of an internal rating system of a bank aiming at rating small and medium enterprises. It should also be compliant with Basel II IRB regulation.

Normally, the unit that is in charge of building the model is immediately involved from the very start of the project, in order to participate in setting objectives and guiding data gathering processes. Instead, in this case study we start from the sole information contained into the dataset collected, assuming that a number of prerequisites are satisfied: the dataset reflects the market segment in which the model will be used, the accounting rules backing data collected are homogenous, borrowers are truly independent (for instance, not belonging to groups).

A few other available pieces of information are: the dataset includes financial statements data and some other borrowers' information; every observation represents a borrower alternatively classified in three classes (performing, substandard, doubtful) indicated by the variable 'status' (and respectively by the codes 3, 2, 1); the last two classes indicate the 'migration' to the class during the 'observation period'; financial statements data are available at 'time zero'; the 'observation period' of borrowers' status is up to one year from time zero.

All remaining information has to be acquired from the dataset contained in the SPSS-PASW file named W_CS_1_OriginalDataSet.sav; therefore, it is the starting point of the analysis. It reflects, on a reduced scale, a real dataset currently used by banks to develop statistical based models for internal rating systems. Open SPSS-PASW Statistics 18.0 and then open the dataset using the following commands:

> *File, Open, Data*; find the directory where W_*CS_1_OriginalDataSet.sav*
> is, select it and then *Open*.

Some basic features of SPSS-PASW deserve to be briefly introduced. Data have been opened in a spreadsheet presenting observations (borrowers) in rows, and variables (borrowers' characteristics) in columns. The bottom left hand side of the screen indicates that the current screen is the 'data view' and there is a second sheet denoted as 'variable view'. Clicking on the second sheet a different view is presented: all variables in the dataset are listed in rows, and each variable is illustrated by a number of attributes:

1. Variable name (name)

2. Data type; numeric, comma for thousands, dot for decimals, scientific notation, date, dollar, custom currency, and string (type)

3. Number of digits or characters (width)

4. Number of decimal places (decimals)

5. Descriptive variable labels; they are more articulated than variable names (label)

6. Descriptive values labels, that is to say meaning of values for categorical variables (values)

7. User-defined missing values (missing)

8. Column width (columns)

9. Alignment (align)

10. Measurement level. Scale is data measured on an interval or ratio scale, where the data values indicate both the order of values and the distance between values. Data with a limited number of distinct values or categories

are said to be categorical variables (also qualitative variables); they can be nominal, when its values represent categories with no intrinsic ranking, or ordinal, when its values represent categories with some intrinsic ranking

11. Role; the variable's role in the analysis (input, target, both, none, partition, split).

All attributes are saved upon saving the data file. You are able to add and delete variables and modify the corresponding attributes of variables. The main menu is identical to that in the 'data view', so that the same analysis can be carried out starting from both windows.

More information can be easily obtained using the extensive Help facilities provided by the SPSS-PASW menu, in particular, Topics and Tutorial.

Another basic aspect of SPSS-PASW is the 'Output File' (with .spv extension), which is automatically opened once you open the dataset file (with .sav extension) and in which a summary of the analyses performed and outputs achieved are reported. The main menu of this file is the same as that of the database file, with the exception of the 'Insert' item, which has been added in order to allow outputs to be edited.

Initially, all variables appear to be numeric and scale. As such, no one associates 'values' to variables. If you scroll vertically in 'variable view' you can also see that:

(a) there are 137 variables;

(b) names of the first four variables suggest that they are part of personal data: BORCODE is the borrower's code assigned by the bank; STATUS indicates the borrower's classification in regulatory grades (performing, substandard, doubtful); SUBSECTOR and SECTOR represent numerical codes for business subsectors and sectors;

(c) following variables are balance sheets and income statement items;

(d) from ROE below, there are a number of derivative variables, indicating flow of funds statement items, financial margins and ratios.

It is evident that the first four variables have been misinterpreted by the procedure used to import data into SPSS-PASW from the bank's information system: they are not scale variables but rather categorical variables and, more specifically, nominal variables. Therefore, we need to change their attributes in the column titled 'measure' of 'variable view'.

Click on the upper cell in the column 'measure' and select 'nominal'. Repeat for the following three variables. Save the file as 'W_CS_1_OriginalDataSet_1 .sav'.

Now we can also define a legend for the variable 'Status'.

In 'variable view', click on the cell at the crossing of row 'status' and column 'values' and, then, on dots, type $value = 1$, $label = doubtful$, add, $value = 2$, $label = substandard$, add, $value = 3$, $label = performing$, add, OK

These labels are now stored in the 'variable view', while the 'data view' still presents the numerical values of the 'status' variable. If you want to view the value labels in the 'data view', you must use the main menu in data view:

View, value labels

Now, doubtful, substandard and performing appear instead of numerical codes in the 'status' column. From the 'data view' window, it is clear that the dataset is ordered by the progressive BORCODE code in the first column. To verify the size and the composition of the dataset in terms of the variable 'status' (performing, doubtful, loss) we can obtain a frequency table by SPSS-PASW using the following commands:

Analyze, Descriptive statistics, Frequencies, select 'Status' and click on the arrow to move to the box to the right, verify that *Display frequency table* is selected, *OK*

The file 'Output1 [Documenti 1]' now presents the result. Save this file.

File, Save as: W_OriginalDatasetOutput, Save

Note that this file contains information at the top concerning which dataset file was used and a script of the carried out commands (saving *W_OriginalDataset-Analysis* file and frequencies analysis on the variable status). It appears as follows:

```
GET
  FILE='C:\Documents and Settings\DeLaurentis\My Documents\AA-
Lavoro\Libro Scoring\W_CS_1_OriginalDataSet_1.sav'.
  SAVE OUTFILE='C:\Documents and Settings\DeLaurentis\My Documents\AA-
Lavoro\Libro '+
  'Scoring\W_CS_1_OriginalDataSet_1.sav'
  /COMPRESSED.
FREQUENCIES VARIABLES=STATUS
  /ORDER=ANALYSIS.
```

Then, results are shown just after the title of the analysis (Table 4.2).

The dataset contains 2515 valid observations, almost 98% of which are performing borrowers. Apart from an observation where the status is unknown, other observations are either doubtful or substandard. The Basel II default definition includes all doubtful and substandard loans, as well as selected past due exposures. It is common in the current transitory period that the bank is unable to identify historical 'past due loans' according to the Basel requirement and, as a consequence, the rating system will be based on a stricter default definition that only includes doubtful and substandard loans (in the model calibration stage, adjustments are made to take into account the regulatory requirement). It follows that in our sample the default rate is 2%. As the observation period lasts one year from time zero, this is a one-year default rate.

Table 4.2   Frequencies.

Statistics		
STATUS		
N	Valid	2515
	Missing	0

STATUS		Frequency	%	Valid %	Cumulative %
Valid	0	1	0.0	0.0	0.0
	doubtful	15	0.6	0.6	0.6
	substandard	36	1.4	1.4	2.1
	performing	2463	97.9	97.9	100.0
	Total	2515	100.0	100.0	

An important feature of SPSS-PASW is that when selecting a routine using the main menu, it is possible to save a script of given commands in a third type of file known as Syntax (having the extension .sps). Once you have selected a routine:

Analyze, Descriptive statistics, Frequencies, select 'status' and click on the arrow to move in the box to the right, verify that Display frequency table is selected.

Instead of clicking on '*OK*', choose '*Paste*'. The syntax file will open automatically and the command script, related to the commands selected, will be in view. In SPSS-PASW release 18.0 the syntax is equal to the information that can be read above the results in the .spv files, but commands listed in .sps files can be activated by selecting 'Run' from the main menu of the syntax program. The script files should be saved, as they can prove to be extremely useful; for example, in a situation where you have to repeat the analysis on different datasets. Save the file as *W_OriginalDatasetSyntax* (the extension is .sps).

Building a borrower rating requires a dichotomous dependent variable that has only two possible states: default/non-default. Therefore, we need to recode the status variable in a new categorical variable in which one category includes both substandard and doubtful cases. From the main menu:

Transform, Recode into different variables, move Status to the box to the right, write Output variable: BadGood, Output Variable Label: 01Status, Change, Old and new values: 1 ->1, add, 2->1, add, 3->0, add, continue, Paste

Do not flag *'Output variables are string'* because for many analyses we need numerical 0/1 values. Now, from the syntax file select the recode syntax and then from the main menu:

Run, Selection

A new variable appears at the right end of the dataset in the 'data view'. Drag it from the last column and drop it to the second column in the dataset, just after the BORCODE code (or cut and paste by right clicking on the column heading). In the 'variable view' we can set the 'variable values' (0 performing, 1 default) for the new variable, as we did previously for the original variable.

We can check that everything is fine by requesting a new frequency table on the new variable by:

Analyze, Descriptive Statistics, Frequencies, BadGood

Alternatively, we can use the syntax file; select the previous command lines regarding the frequency table obtained previously, for the original 'status' variable, and copy and paste after the Recode commands; now we just need to change the label 'status' with 'BadGood' and run the last commands.

The default rate is 2%, and there is a missing value (as 'status' has the unexpected value of zero which we have not recoded into any of the new variable's values). In order to identify the observation with the unexpected status code, we can go into the 'data view' window, right-click on the heading of the column 'status' and choose to arrange this variable in 'ascending order'. The first row will present the observation with status code zero. It is obvious that this is a very odd observation with all variables set to zero. This means that the dataset still requires some refinement.

This observation can be easily erased by clicking on the row heading '1' (in order to select the entire row) and then using the 'canc' button. In *W_Original-DatasetOutput.spv* file we can take into account the modification that was made to the dataset; from the main menu:

Insert, New text, Variable with BORCODE 2476 has been erased because all variable's values were zero,
File save

Moreover, the new dataset in the .sav file should be saved (as *W_CS_1_ OriginalDataSet_2.sav*) after having rearranged the first column (BORCODE) in ascending order.

We can now gain some other insights regarding the dataset contents by examining the distribution of the two other categorical variables SUBSECTOR and SECTOR. Having the newly created dichotomous variable BadGood, it would be of interest to obtain the distribution separately for the two subsets of defaulted and non-defaulted firms: in doing so, it is possible to calculate whether there is a similar distribution of bad and good firms per business and economic sectors or, if the two subsets behave differently.

We must first select the following from the main menu:

Data, Split file, Compare groups, 01Status, Paste

Then:

Analyze, Descriptive statistics, Frequencies, select 'SUBSECTOR' and 'SECTOR', click on the arrow to move in the box to the right, verify that Display frequency table is selected, Paste

And eventually, *Run* the previous commands in the syntax file. It is evident from the output file that SUBSECTOR is too detailed and very few observations are available for many businesses, whereas by using SECTOR observations are far more aggregated. Nevertheless, from both perspectives distributions of bad and good cases are dissimilar and, also, SECTOR presents many categories which are insufficiently populated.

The two tables obtained can be rearranged in order to improve the comparability between the two subsets.

Double click on a table, select Pivoting trays, drag 01Status from the left side of the window and drop it on the upper side just below 'statistics'. File, Close

Then, repeat the process for the other table and finally save all opened files. Note that when you save SPSS-PASW files that have had further changes made since the previous save, the asterisk will disappear from the file name in the bottom taskbar of Windows.

Two other important checks and refinements concern the possible presence of duplicate cases and missing values.

4.3.2 Duplicate cases analysis

Preliminary analysis of the dataset and data cleansing are key steps to avoid extending to credit risk models the well know rule 'garbage in, garbage out'. There are a number of reasons why the same borrower may be included twice or more in the dataset. In order to have independent observations, this situation must be solved by erasing any duplicates. Of course, basic analyses are carried out when gathering data from the banks' information systems; as a result, duplicates do not usually have the same BORCODE but, nevertheless, when other variables' values are equal we can argue that apparently different observations are related to a unique firm.

SPSS-PASW has a specific function which allows us to spot duplicate cases by using a list of variables we can select. Let us state that the observations sharing the same Total assets, Inventories, Equity, and Revenues are actually related to the same company. Reconsider the *W_CS_1_OriginalDataSet_2.sav* dataset. If *Split File* is still on, from the main menu:

> *Data, Split file, Analyze all cases. Do not create groups*

Then:

> *Data, Identify duplicate cases, Define matching cases* by*: Net Inventories, Total Assets, Total Stockholders' Equity, Sales,* flag *Move duplicate case to the top of file,* flag *Sequential count of matching cases in each group*

The output file shows that there are four duplicate cases, whereas 2510 are primary cases. If we select 'data view', we can identify the four observations considered to be duplicate cases, which are located at the top. They have different BORCODE values, but they are clearly the same firms. BORCODE 1371 and 1393 have almost all the same values for each variable, but a different SUBSECTOR is shown. BORCODE values 1436 and 1439 have slight differences in some variables; for instance, gross fixed assets and cumulated depreciations are distinctively indicated for the former observation, whereas the latter has the net fixed assets indicated as gross and zero cumulated depreciation. BORCODE values 151 and 152 differ for SUBSECTOR and SECTOR alone. BORCODE values 954 and 955 present minor differences in some financial indicators.

In the last column of 'data view' a new variable has been added. Erase the results indicated as 'Duplicate case', then sort the dataset by BORCODE (ascending) and save it as:

> *W_CS_1_OriginalDataSet_3.sav.*

Finally, save the output and the syntax files.

4.3.3   Missing values analysis

Missing value analysis is executed in order to verify dataset completeness and to decide how to deal with missing data. It is necessary to handle missing values because: (i) in mainstream multivariate procedures, values must always be available for each indicator, otherwise it is not possible to get a rating; (ii) missing values could derive from causes that reduce the representativeness of data. Consequently, quantitative relevance and causes of missing values must be carefully looked after.

The main causes for missing values in datasets are:

(a) not applicable (for instance, in the case of Inventories for a service company),

(b) not available (uncollected; it is typical of SECTOR code),

(c) unknown (this is usually the case when questions are not answered in questionnaires),

(d) value zero of the variable (true zero).

To identify missing values, from the main menu:

Analyze, Missing value analysis, Use all variables, move from the upper box to the lower box: the first five variables that are categorical, *Paste*

In the syntax file, find the rows concerning MVA (Missing Values Analysis):

change */MAXCAT* = 25 to */MAXCAT* = 2600, then select MVA commands lines and *Run Selection*

In the Output viewer, MVA indicates that there are only two missing values, concerning SUBSECTOR and SECTOR. Therefore, we have a negligible problem arising from this MVA.

In the 'data view' window we can right click on the SUBSECTOR column heading and select *Sort ascending* for the variable. We can see that the two missing values concern the same observation (BORCODE 1018). In real situations, we can either ask bank's officers to retrieve the information or we can erase the observation (as it is a performing borrower, we are not going to reduce the small number of defaults in our dataset). Here we opt to erase it.

A quick glance at the sorted dataset outlines that there are two dozen zeros for SUBSECTOR. This is an example of a larger problem concerning false non-missing values.

In fact, numerical codes that banks use to indicate absence of data or abnormal situations should also be treated as missing values: they do not represent variables values, instead they have particular meanings. Often a complete list of all codes used in banks' datasets is not available, therefore it is necessary to identify these false non-missing values checking whether the frequency of values such as 0, 9999, 99999 is not normal.

To perform this new missing value analysis, we need to define which numerical values may possibly be hiding missing data; by checking their frequency we can work out if the value occurs by chance or if it is in fact a code of some sort. It is important that we carry out the analysis separately for different types of values, in order to gain specific answers for each of them. Start with the zero values.

In 'variable view' there is a column titled 'Missing' where we can set the values we would like to check for. Choose a cell, for instance SUBSECTOR, *Discrete missing value,* digit 0, *OK.* Now right-click on the cell and *Copy* it. Then select all other cells in the 'Missing' column with your mouse, up until the last financial indicator (the row before the last), and *Paste.* Save the dataset as *W_CS_1_OriginalDataSet_4x.sav.*

Now, access the syntax file, select the same MVA command lines used before and *select Run Selection.* In the 'output viewer' we can observe that many variables present a vast number of zeros. They may be completely normal for many financial statement items related to specific assets, liabilities, costs, and revenues. They are abnormal for summary items such as Total Assets (12 cases) or Value of Production (22 cases). More importantly, we should check missing values for financial ratios as they are the natural candidates to input into the scoring model. In fact, in order to avoid multiple dependencies on firms' size (which, in case, may be considered a variable by itself) by different explanatory variables, it is more suitable to use ratios instead of other types of expressions of financial performance. If we look at the missing values for ratios, we can see that:

(a) There were no missing values in the previous analysis regarding the absence of data; this is clearly odd because it means that the denominators were never zero. More probably, the explanation for this is that during the data gathering processes that situation was managed by forcing the ratio to zero.

(b) In the new analysis concerning zeros, the highest percentage of missing values (about 23%) is linked to various ratios involving flow of funds (CashFlowFromOperations/Sales, CashFlowFromOperations/ValueOfProduction, CashFlowFromOperations/EBITDA, CashFlowFromEarnings/Net FinancialDebt). Evidently, the flow of funds statement was often lacking in the original bank's dataset.

(c) The second highest percentage of missing values (about 17%) concerns ratios involving inventories. They may clearly be zero for firms operating in the service sector, so they are false-missing values.

(d) There is approximately 10% of missing values for the cost of financial debt, which is possibly due to firms with no debt. 8% is missing values for receivables divided by total assets, not including the 12 cases with zero total assets that we must investigate further. The remaining cases could possibly be related to firms selling for cash. About 5% is the percentage of missing values for some ratios concerning payables, this could actually be zero for some firms which pay cash and take discounts; so they are all false-missing values.

(e) Other financial ratios present quite a low percentage of missing values, so they can be easily managed by substitution approaches.

The last step in missing analysis requires spotting numerical codes such as 999 and the like, which could be dramatically misinterpreted as true values. To carry out the analysis it is adequate to replicate the same process used to spot zeros using, for instance, 999, 9999, and 9999 as missing values (Figure 4.2). We can use exactly the same commands listed in the syntax file for MVA. Looking at the output viewer, results are encouraging because of the very low number of cases matching the criteria: it is clearly a matter of chance and they can be considered true values.

We can now save the files *W_OriginalDatasetOutput.spv*, *W_OriginalDataset-Syntax.sps* and the dataset as *W_CS_1_OriginalDataSet_4xx.sav*. We have added

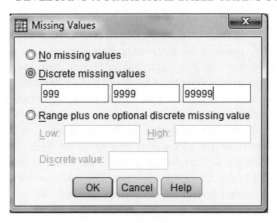

Figure 4.2   Missing value analysis.

x and xx to the files names when saving the last two datasets because zeros for the former and 999, 9999 and 99999 for the latter have been defined (and saved) for all variables as missing values. Consequently, we must remember that if we are going to use these files for future analysis, all those values would not be treated as true values but as missing. On the contrary, we have already clarified that there are no problems with 'type-9-codes' and we must still decide how to deal with zeros.

4.3.4   Missing value treatment

The missing value analysis has tried to identify the 'true' missing values (including false non-missing and excluding false missing that are real values). The management of missing values depends on their quantity and their concentration:

1. If missing values are abundant in specific observations, it may be necessary to exclude the observation. If it is a defaulted firm, due to the often low number of defaults, it could be convenient to collect data manually to fill in missing values.

2. If missing values are abundant in specific indicators, it could be necessary to exclude such indicators from the dataset.

3. If many indicators present many missing values, it may be necessary to transform continuous variables into categorical ones, one category being that of missing values. Due to the nominal nature of this type of transformed data, it is indeed possible to estimate a connection between a value which cannot actually be determined and its effects on creditworthiness. Of course, categorical variables produce a weaker signal for multivariate procedures, so this solution is a last resort.

4. If a few indicators present a small percentage of missing values, we can choose to either not treat them at all or to replace them with the mean or

median values for the indicator. Median is preferred to mean because the latter can be strongly impacted by outliers. An ongoing debate concerns the dataset to consider for calculating these central tendency measures: the overall dataset or the two separated groups of bad and good cases? An interesting position is stated by OeNB and FMA (2004): 'group-specific estimates are essential in the analysis sample because the indicator's univariate discriminatory power cannot be analyzed optimally in the overall scoring function without such groupings. In the validation sample, the median of the indicator value for all cases is applied as the general estimate. As the validation sample is intended to simulate the data to be assessed with the rating model in the future, the corresponding estimate does not differentiate between good and bad cases. If a group-specific estimate were also used here, the discriminatory power of the resulting rating model could easily be overestimated using unknown data'. In the event that missing values will be substituted by mean or median values in order to estimate a model, it is necessary to perform all other univariate and bivariate analysis of indicators before handling missing values to avoid reaching the wrong conclusions, for instance, on their discriminatory power.

The lower the fraction of valid values for an indicator or a case in a dataset, the less suitable the indicator or the case is for developing a rating model; it is useful to set a clear policy, defining a minimum level of valid cases up to which availability for an indicator or a case is considered sufficient. This is also true for zeros, either when they indicate a true numerical value for the variable and when they indicate that data is absent.

A possible policy is:

(a) Observations are erased if any key data (such as Total Assets or Value of Production) or more than 75% of indicators' values are missing.

(b) Indicators are erased if they present more than 10% missing values.

(c) Median is substituted to missing values if the indicator has less than 10% of missing values. Median has to be calculated separately for Bad and Good groups for the analysis sample, and as a single value for the validation sample. Consequently, the substitution has to be done after splitting the dataset into the two samples. Specifically, substitution will also follow all univariate and bivariate analyses, in order to avoid improper influence on them.

(d) Indicators are transformed into categorical values if they are considered relevant and if there are only a few other variables in the dataset.

Of course, there are interconnections that exist among the four statements mentioned above. For instance, if we first erase indicators and then check for the percentage of non-missing values for observations, the end result can be different than if we apply rules in reverse order. The model builder should verify case by case which order is the best to use.

Applying that policy to *W_CS_1_OriginalDataSet_3.sav* means:

(a) Erasing cases with Total Assets (12 cases) or Value of Production (22 cases) equal to zero. We must access the 'data view', sort the dataset by ascending order regarding one of these variables at a time, select observations with zeros for the variable and delete the entire rows using the 'del' button on the keyboard. Before doing so, it would be of interest to examine the accounts of these odd firms: those with zero Total Assets usually have Gross Fixed Assets completely compensated by the Accumulated Depreciation and no other items on the asset side; all cases having a Value of Production equal to zero also have Net Sales equal to zero.

(b) Erasing the four flow of funds indicators (CFfromOPERATIONSon-SALES, CFfromOPERATIONSonVP, CFfromOPERATIONSonEBITDA, CFfromEonNET_FINANC_DEBTS) which have about 23% of zeros. From the 'variable view', whilst holding the 'Ctrl' button, using the mouse select the four rows containing these variables and press the delete button on the keyboard.

(c) Taking ratios involving inventories without changes, considering the 17% of zeros as false-missing values (firms can possibly operate in the service sector).

(d) Be ready to substitute the median to all other financial ratios once the analysis sample and the validation sample are created.

After carrying out these procedures, sort the dataset in ascending order, according to BORCODE and save the dataset as *W_CS_1_OriginalDataSet_5.sav*.

4.3.5   Other preliminary overviews

There are a number of other preliminary data explorations which can be performed before moving on to create the analysis sample and the validation sample. As an example, we are going to check consistency of data in terms of firms' size. This can greatly affect the structure of accounting data and the path to failure that scoring models attempt to identify. A basic analysis on Sales can be achieved by (Figure 4.3):

> *Analyze, Descriptive statistics, Explore, Dependent list: NetSales, Display: Statistics*

Then, click on Statistics on the upper right hand corner, and select *Descriptives* and *Percentiles*. Click on *Paste,* in order to obtain the syntax. Once you run the syntax of the Explore procedure, problematic results are obtained, which are shown in Table 4.3.

Worries derive from high level of skewness of the distribution of Sales values, not only expressed by the proper statistic, but also quite simply by the

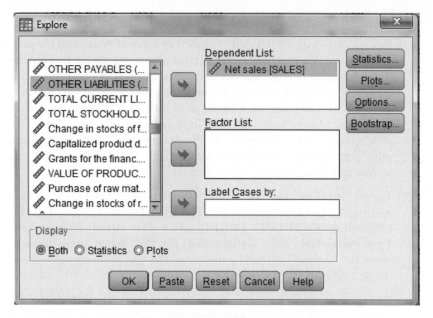

Figure 4.3   Explore.

huge difference among the mean, the trimmed mean, and the median. Also, the great distance between the 95th percentile and the maximum value indicates that there are a few observations with much larger Sales than the great majority of other firms.

These results push for a more direct analysis of the dataset. In the 'data view', sorting (in descending order) by NetSales, we see a clearer picture. Only seven firms have net sales above Euro 500 000 thousand. These are large companies in terms of sales, total assets, net worth, and the like. The numerical weight in the dataset is trivial but they are clear outliers in terms of size. It is advisable to exclude these companies from the dataset, something which we do. Then, save as *W_CS_1_OriginalDataSet_6.sav*.

Before running the syntax again for Explore routine, copy and paste it below in the syntax file, then add 96, 97, 98, 99 among percentiles as in the script below:

```
EXAMINE VARIABLES=SALES
/PLOT NONE
/PERCENTILES(5,10,25,50,75,90,95, 96, 97, 98, 99) HAVERAGE
/STATISTICS DESCRIPTIVES
/CINTERVAL 95
/MISSING LISTWISE
/NOTOTAL.
```

Then, *Run.* Results in Table 4.4 show a strong improvement in the consistency of data, obtained at the expense of only seven observations, all belonging to the wide group of performing borrowers. The skewness is much lower and the

Table 4.3   Explore routine output.

	Case processing summary					
	Cases					
	Valid		Missing		Total	
	N	%	N	%	N	%
Net sales	2,476.0	100.0	0.0	0.0	2,476.0	100.0

Descriptives			Statistic	Std. Error
Net sales	Mean		29,438,477.64	19,120,928.86
	95% Confidence	Lower Bound	−8,056,190.37	
	Interval for Mean	Upper Bound	66,933,145.65	
	5% Trimmed Mean		4,212.65	
	Median		1,959.50	
	Variance		9.05 E 17	
	Std. Deviation		9.51 E 8	
	Minimum		.00	
	Maximum		42,214,978,215.00	
	Range		42,214,978,215.00	
	Interquartile Range		4,992.00	
	Skewness		39.14	0.049
	Kurtosis		1,641.04	0.098

Percentiles								
		Percentiles						
		5	10	25	50	75	90	95
Weighted Average (Definition 1)	Net sales	168.85	315.70	756.75	1,959.50	5,748.75	15,852.80	31,248.60
Tukey's Hinges	Net sales			757.50	1,959.50	5,747.50		

Table 4.4   A new Explore routine output.

Case processing summary

	Cases					
	Valid		Missing		Total	
	N	%	N	%	N	%
Net sales	2,469	100.0	0.0	0.0	2,469	100.0

Descriptives

		Statistic	Std. Error
Net sales	Mean	7,411.77	423.09
	95% Confidence Interval for Mean Lower Bound	6,582.12	
	Upper Bound	8,241.42	
	5% Trimmed Mean	4,131.97	
	Median	1,955.00	
	Variance	441,968,005.93	
	Std. Deviation	21,023.04	
	Minimum	.00	
	Maximum	464,186.00	
	Range	464,186.00	
	Interquartile Range	4,970.00	
	Skewness	9.80	0.049
	Kurtosis	149.18	0.098

Percentiles

		5	10	25	50	75	90	95	96	97	98	99
Weighted Average (Definition 1)	Net sales	168.50	315.00	755.50	1,955.00	5,725.50	15,472.00	30,788.50	38,223.20	49,217.30	61,663.00	85,910.70
Tukey's Hinges	Net sales			756.00	1,955.00	5,721.00						

previously larger noted differences among the mean, the trimmed mean, and the median have been greatly reduced. Also, NetSales values at various percentiles beyond the 95th are considerably closer.

4.4   Defining an analysis sample

4.4.1   Rationale for splitting the dataset into an analysis sample and a validation sample

Data gathered and cleansed in the previous stages represent the overall dataset. However, we need data to estimate the model and different data to validate it by a hold-out sample, that is to say, through data that have not contributed to the model's estimation. This is the most common and robust test for observing the possibility of generalizing the model's discriminatory power to cases different from those by which the model was built. This is also a typical situation in which the model would be used for in daily operations.

Consequently, the overall dataset has to be divided into an analysis sample and a validation sample.

The 'analysis sample' supports the development of the scoring model, whereas the 'validation sample' serves to quantify the (probable) deterioration in the model's performance moving from in-the-sample to out-of-sample measures. Of course, borrowers included in the analysis sample must not be used in the validation sample, independently from the possible different time frame of data, since the deep idiosyncratic characteristics of the borrower would indeed remain the same.

When the dataset size is small, it may be necessary to use alternative 'cross validation' approaches to the actual splitting of the overall dataset into two samples. The 'bootstrapping method' iterates the estimation of the model several times, while each time it excludes a certain (small) number of cases (even just one case) from the analysis sample which are randomly chosen and will be used to validate each model out-of-sample. The mean and variability of the models' performances serve as indicators of the overall scoring function's discriminatory power. Modern high speed computers, even personal computers, have solved the problem of computational requirements for this approach. SPSS-PASW provides a 'leave-one-out classification' facility that is easy to generate while developing discriminant analysis, as we will see.

4.4.2   How to split the dataset into an analysis sample and a validation sample

The actual division of the dataset into analysis and validation samples is the preferred option as long as sufficient data are available. To avoid any bias due to subjective division, a random selection process should be used. After splitting the sample, it would be convenient to verify that the composition of the two samples by relevant dimensions of borrowers (such as size, sectors, geographical areas, calendar year of 'time zero') is not too dissimilar.

If data were gathered as a 'full survey' of the target segment of the bank's portfolio, the proportion of bad cases is representative of the segment to be analyzed. When splitting the dataset by a random procedure, the same proportion should be obtained. However, when the default rate of the target segment is low, it is favorable to increase the proportion of bad cases in the analysis sample.

Consider that:

(a) in the pioneering studies on credit scoring models development, it was commonplace to have a fifty-fifty proportion of good to bad cases, in order to have equal prior probabilities;

(b) some researchers suggest that approximately one quarter to one third of the analysis sample comprises of bad cases (OeNB and FMA, 2004);

(c) others prefer to mirror the actual default rate of the portfolio target segment.

A very low proportion of default cases may hamper the reliability with which the statistical procedure can identify the differences between good and bad borrowers. It is not simple to define a priori the minimum acceptable sample size, above all for defaulted observations. For instance, once the model has been estimated and rating classes defined, in order to quantify the default risk of rating classes some statistical tests require a minimum number of five cases of default per class. However, meeting this requirement depends on the number of rating classes created and on the per class distribution of defaults (which will be only known at the end of the model development process).

In the current case study, we prefer to increase the percentage of defaults in the analysis sample, using all available defaults and only approximately half of the performing borrowers. This choice does not solve all issues concerning the minimum number of defaults (this is re-evaluated more in- depth in Chapter 5) and has two consequences: (a) defaulted borrowers will be part of both the analysis and the validation samples; (b) the model's calibration will require the rescaling of estimated default probabilities. Therefore, we must select all bad cases and store them in a new dataset. The easiest way to do this would be to:

open *W_CS_1_OriginalDataSet_6.sav,* order data by descending order of the variable 'BadGood' (labelled 01Status), delete all observations classified as 'performing' and save the residual dataset as *'BadSample.sav'*

For good cases, we would like a random selection of 50% of these cases. We can:

open *W_CS_1_OriginalDataSet_6.sav.,* order data by descending order of the variable 'BadGood', delete all observations classified as 'default' and save the residual dataset as *'GoodSample.sav'*

SPSS-PASW main menu can now provide the random selection tool:

> *Data, Select cases, Random sample of cases, Sample, Approximately 50% of cases, Continue, Delete unselected cases, OK, Save as 'RandomGoodSample.sav'*

Now, with all SPSS-PASW files closed except *BadSample.sav* and *RandomGoodSample.sav*, from the main menu of one of the two datasets, request the following commands:

> *Data, Merge Files, Add cases, Select the other opened. sav file, Continue, OK*

Now we have the Analysis Sample we were looking for. Erase the last variable 'PrimaryLast' that was created during the duplicate cases procedure and then save the new dataset with a name of your choice. Effectively, as the randomization procedure would produce different results each time it is applied, we are going to use a given analysis sample, which can be found under the name W_CS_1_AnalysisSampleDataSet.sav.

Now, request summary statistics of the composition of the analysis sample in terms of the variable '01Status'. We can request a frequency table by:

> *Analyze, Descriptive statistics, Frequencies,* select *01Status* and click on the arrow to move to the box on the right, verify that *Display frequency table* is selected, *Paste*

From the new syntax file choose:

Run all
File, Save as: W_AnalysisSampleSyntax_1, Save

The file 'Output1 [Document 1]' now presents the result. Save this file:

File, Save as: W_AnalysisSampleOutput_1, Save

The number of defaults in the sample is 51, whereas there are 1221 performing borrowers, for an overall number of 1272 observations and a sample default rate of 4%, which is, as desired, about the double of the default rate we had at start. Now we are all set for a comprehensive analysis of the usefulness of potential explanatory variables included in the sample.

4.5   Univariate and bivariate analyses

The purposes of univariate and bivariate analyses are to identify creditworthiness characteristics which have an economic meaning and related data suitable for multivariate statistical procedures. The end result of this stage is a shortlist of independent variables for the ensuing multivariate analyses. These two requirements (economic meaning and statistical aptness) give practical and theoretical consistency to subsequent analyses and, at the same time, reduce their complexity.

Univariate analysis considers one indicator at a time, possibly separating two subsamples for defaulted and non-defaulted observations and comparing their central tendency values and distributions.

Bivariate analysis compares either two indicators or an indicator (independent variable) with the dependent variable (the dichotomous status default/non-default).

4.5.1 Indicators' economic meanings, working hypotheses and structural monotonicity

4.5.1.1 Scale variables

Consider financial ratios that, as mentioned, are preferred to pure financial statements items because the latter are relative values depending on the firm's size. Ratios are always numerical scale variables. There are 30 financial ratios in the dataset; these are ratios between financial statements data concerning a number of different economic and financial characteristics of borrowers.

The first two tasks to undertake on such ratios are:

1. understanding their economic meaning and set a working hypothesis on the sign of the expected relation with probabilities of default (PDs);

2. check if ratios are structurally monotonic in regards to default risk (a short way of saying that the ratio structure permits the relation between a ratio's values and default risk to be monotonic).

An indicator is considered a reliable predictor of default risk if it behaves empirically according to economic theory. In this case, we can assume that it is not only correlated by chance with default risk, but it also summarizes facts that have a substantial economic connection with default probability. The test requires a statement in regards to a working hypothesis that describes the sign of the relationship with the probability of default, due to economic reasons. For example, for ROE, we can state a negative relationship with PD; in fact, a more profitable firm – where all other conditions are equal – is less risky because profit:

• can be retained to increase capital,

• will sooner or later become cash,

• makes it easier to borrow new money or collect new equity for a firm that can assure a higher payout.

The working hypothesis can be verified on the basis of the analysis sample by comparing means and/or medians of the two subsamples of good and bad borrowers (quite clearly in our example we do expect that they are higher for the subsample of performing firms). Only when an indicator's working hypothesis can be set and can be empirically confirmed it is possible to use the indicator in further analyses.

If financial theory is contradicted by empirical results based on the analysis sample, the indicator should be excluded from the short list of possible explanatory variables being entered into a multivariate analysis. If financial theory does not provide a clear and unidirectional working hypothesis whereas empirical tests show a strong and statistically significant relationship with the default rate, the indicator should be treated with caution; there is, in all probability, a secondary relationship to search for that may explain the connection.

For scale variables such as financial ratios, a second analysis, which can be performed immediately, is to verify if we can assume the existence of a monotonic relationship with default risk. This is a crucial requirement for mainstream statistical multivariate approaches to rating models such as logistic regression and discriminant analysis (whereas artificial neural networks and decision trees, for instance, do not need it). There are two types of monotonicity: structural and empirical.

In mathematics, a monotonic function $f(x)$ is a function which preserves the given order of x; for instance: if for all x and y such that $x \leq y$, we have $f(x) \leq f(y)$ we say that the function is monotonically increasing (if, whenever $x \leq y$, then $f(x) \geq f(y)$, the function is named monotonically decreasing). We define a financial ratio as structurally monotonic when the range of variability of the denominator theoretically assures that the ratio always moves in the same direction, consequently preserving a unidirectional (positive or negative) relationship with the risk of default. Therefore, structural monotonicity can be studied by examining denominators of ratios and their range of variability: x is the denominator; $f(x)$ is the financial ratio. If financial ratio values always move in the same direction when a denominator increases or decreases, then the ratio is structurally monotonic. As we will see, problems arise when denominators change sign, thereby changing the direction of $f(x)$.

The condition of empirical monotonicity is met when, categorizing the variable into contiguous ordered intervals and calculating the average default rates of borrowers included into such intervals, we obtain a monotonic increase (or decrease) of default rates.

For now, we only have to check the structural monotonicity. Empirical monotonicity will be analyzed later and, as we shall see, it can contradict conclusions reached when examining structural monotonicity. In order to use indicators which do not conform to monotonic working hypotheses, it is necessary to transform these indicators. Sometimes it is easy to solve the issue of structural non-monotonicity by changing the way the ratio has been built (i.e., its structure); in other instances, it is necessary to use more complex solutions that can, however, generally solve both the structural and empirical monotonicity problems of indicators (these solutions are discussed later).

Four aspects have been outlined for each financial indicator in the dataset:

(a) their economic meaning,

(b) the working hypothesis concerning their expected relation with the probability of default,

(c) an evaluation of their structural monotonicity,

(d) in the case of structural non-monotonicity, a possible solution.

This analysis will permit us to fill in the first column and a part of the second column in Table 4.5. The table has been filled in for the whole row for ROE as an example.

Now, let us examine one by one the 30 ratios.

1. *ROE:* it represents the return on equity, that is to say, the accounting profitability of the firm. A negative relation with the probability of default is expected: the higher the ROE, the lower the default risk. The ratio is calculated as net profit divided by equity and is represented as a percentage. The denominator represents the total stockholders' equity net of profit or loss of the financial year, gross of dividends to be distributed, and net of loans extended by the firm to its shareholders. The deduction of loans extended to shareholders is a common practice for SMEs and intends to take into account the resources exhausted by owners of the firm. However, this practice appears to be robust in the event of a judgmental-based credit analysis, but it creates problems in case of statistical-based approaches because the range of variability of the ROE denominator includes negative values: this determines a non-monotonic relationship with the default risk, as shown in Figure 4.4. When the denominator (DEN) decreases – while being positive – ROE increases and default risk lowers; but if denominators decrease further, where its value becomes negative, ROE decreases and alters the relationship with default risk. Note that once the denominator is in a negative region, the lower it is, the higher the ROE. Return on equity relationship with default risk is structurally non-monotonic.

In addition, the ROE numerator can also be negative: the result is that a firm that produces losses and has equity lower than loans extended to shareholders has a positive ROE.

A possible solution is to calculate the 'traditional ROE', without subtracting loans to shareholders from equity.

Lastly, we can argue that the traditional ROE on the one hand is structurally monotonic and, on the other hand, it is probably not empirically monotonic, because a high ROE value can be derived from both high profit and from low capital. This last case can occur frequently for firms that have suffered losses for years, have eroded their capital basis, and have a high default risk. Therefore, a more complex transformation is required; it will be discussed later in this chapter.

2. *EBITDA/Sales:* data is shown as a percentage. It indicates a gross return on sales; a negative relationship with the probability of default is expected. Denominator is necessarily positive: the indicator is structurally monotonic with respect to default risk.

3. *ROI:* EBIT divided by total assets, shown as a percentage. It indicates the operating profitability of operating assets. A negative relationship with the probability of default is expected. Denominator is necessarily positive: the indicator is structurally monotonic with respect to default risk.

Table 4.5 Summary table of univariate and bivariate analyses: an example.

# Ratio	Structural and Empirical Monotonicity: It has economic meaning (Y/N); sign of relation with PD (+, −)	Means. Trimmed means and Medians are aligned with hypotheses	Histograms and Q–Q plot suggest normality of distributions Good Bad subsamples	Normality test (K–S; SW) is not significant (p-value >0.05) Good Bad subsamples	Homogeneity of Variance (Levine's) test is not significant (p-value >0.05)	Means difference t-test is significant (p-value <0.05)	Means difference F-test is significant (p-value <0.05)	F-value	AuROC value Statistical significance at 5%	Correlation >70% Pearson's *Spearman's*	# Outliers (3 × IQR)	Conclusions
Legend	Y/N + −	Y/N Y/N Y/N Y/N	Y/N Y/N	Y/N Y/N	Y/N	Y/N	Y/N	Value	Value Y/N	Ratios	Number	
1 ROE	Y (−)	N N Y Y/N	N N	N N N N	N	N	Y	14.9	59 Y	ROEtr	154	
2 EBITDA on SALES	N N	N N N	N N	N N	N	N						
3 ROI												

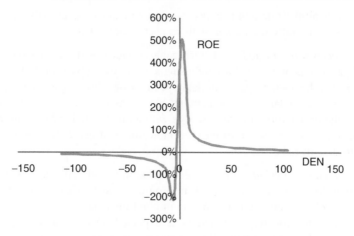

Figure 4.4   ROE as a function of its denominator.

4. **ROA:** current income divided by total assets, shown as a percentage. It indicates the profitability of assets before interest expenses. A negative relationship with the probability of default is expected. Denominator is necessarily positive: the indicator is structurally monotonic with respect to default risk.

5. **EBITDA/ValueOfProduction:** a negative relationship with the probability of default is expected. A negative denominator is rare; therefore, we can consider the indicator as structurally monotonic with respect to default risk. In order to check the frequency of negative denominators, we can order the analysis sample by selecting *Sort Ascending* on the variable value of production; only two cases (both performing borrowers) have negative values.

6. **ROS:** EBIT divided by sales, shown as a percentage. It indicates the ratio of earnings before interest and taxes to sales, that is to say, the operating profitability of sales. A negative relationship with the probability of default is expected. Denominator is necessarily positive: the indicator is structurally monotonic with respect to default risk.

7. **Assets_Turnover:** sales divided by total assets. It indicates how many times a year the firm is able to sell the (value of) assets; it is an expression of the efficiency of assets in creating sales. A negative relationship with the probability of default is expected. Denominator is necessarily positive: the indicator is structurally monotonic with respect to default risk.

8. **Inventory_Turnover:** sales divided by inventories. It indicates how many times a year the firm is able to sell the (value of) inventories. A negative relationship with the probability of default is expected. Denominator is necessarily positive: the indicator is structurally monotonic with respect to default risk.

9. **Receivables_Turnover:** sales divided by trade receivables at year-end. It indicates the capability of the firm to cash in its receivables. A negative relationship

with the probability of default is expected. Denominator is necessarily positive: the indicator is structurally monotonic with respect to default risk.

10. *Receivables_Period:* trade receivables divided by daily sales (net of accumulated provisions on receivables and using a conventional year of 360 days). This ratio indicates the average payment delay in favor of a firm's customers. A higher ratio presupposes a larger investment in current assets for each Euro of sales, therefore, the probability of default moves with a direct relationship with the ratio. Denominator is necessarily positive: the indicator is structurally monotonic with respect to default risk.

11. *Inventory_Period:* ratio between inventories and daily sales. It is a measure of how long raw materials, work in progress, and finished goods stay in the warehouse and, as a consequence, of the implied financial needs. A positive relationship with the probability of default is expected. Denominator is necessarily positive: the indicator is structurally monotonic with respect to default risk.

12. *Payables_Period:* trade payables divided by daily purchases. It estimates the average delay in days by which the firm pays its suppliers. A negative relation with default risk is expected, because high values of the ratio indicate that, at each point in time, more financial resources can be used to finance assets rather than to pay suppliers, and that the firm has a larger contractual power against them. Actually, high values of the ratio may also indicate that the firm has financial difficulties and it tries to shift them onto suppliers. However, in financial analysis, the former argument prevails on the latter. Denominator is necessarily positive: the indicator is structurally monotonic with respect to default risk.

13. *Commercial_Wc_Period:* trade receivables plus inventories minus trade payables, divided by daily sales. It indicates the period between paying out purchases and cashing in sales; this period shows how long the operating cycle of the firm takes (in days). The higher this is, the higher the financial exposure of the firm created by business operations is. A positive relationship with the probability of default is expected. Denominator is necessarily positive: the indicator is structurally monotonic with respect to default risk.

14. *IEonEBITDA:* interest expenses divided by EBITDA, shown as a percentage. It indicates the share of gross operating earnings necessary to cover interest expenses. A positive relationship with the probability of default is expected. Denominator is not necessarily positive: this determines the structural non-monotonicity of the indicator. For this ratio the solution is straightforward: its inverse is structurally monotonic as interest expenses can only be positive (Figure 4.5).

15. *NIEonEBITDA:* net interest expenses divided by EBITDA, shown as a percentage. Here, interest expenses include financial losses and written downs and are net of financial income. A positive relationship with the probability of default is expected. Both denominator and numerator are not necessarily positive: this determines the structural non-monotonicity of the indicator and the impossibility

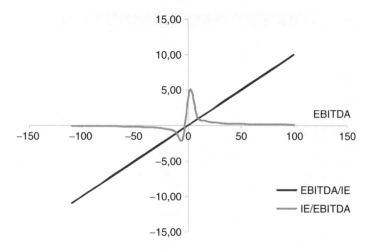

Figure 4.5   Ratios values as a function of EBITDA.

to adopt the simple solution outlined for the previous ratio. In order to check the frequency of occurrence of a negative numerator in the Analysis Sample, a new variable is needed. From the SPSS-PASW main menu (Figure 4.6):

> *Transform, Compute variable, Variable name:* for instance IEminus-FININCOMplusFINLOSWD, and in *Numeric Expression* let's indicate the sum we need, *OK*

The new variable has been added to the right end of the dataset; by selecting *sort ascending* we discover that there are 51 cases with a negative figure (all performing borrowers). Given that the phenomenon is not trivial in terms of frequency, we cannot adopt the simple transformation used for the previous ratio and have to conclude that the indicator is not structurally monotonic. Now, save the new dataset file:

> *File; Save as: W_CS_1_AnalysisSampleDataSet_1x.sav, OK*

16. *IEonLiablities:* interest expenses divided by liabilities, shown as a percentage. It represents the average cost of liabilities. A positive relationship with the probability of default is expected. Denominator is necessarily positive: the indicator is structurally monotonic with respect to default risk.

17. *IEonFinancial_Debts:* interest expenses divided by financial debts, shown as a percentage. This is a more specific indicator than the previous one and indicates the average cost of financial debts. A positive relationship with the probability of default is expected for many reasons: lenders require higher interest rates to more risky borrowers, the firm has a higher cost to bear, financial leverage is riskier (all other variables being equal) because the spread between

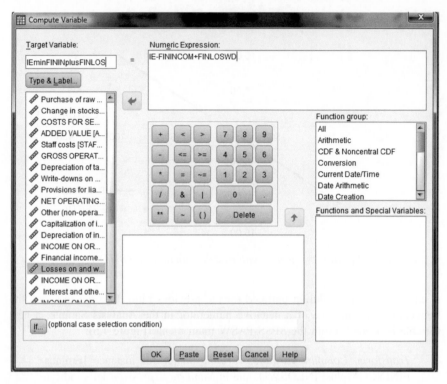

Figure 4.6   Transforming NIEonEBITDA.

ROA and cost of debt is lower. Denominator is necessarily positive: the indicator is structurally monotonic with respect to default risk.

18. ***Extr_I_C_onCurrent_Income:*** It indicates the impact of extraordinary income and cost on current income and it can be calculated as gross of tax profit divided by income from ordinary (current) activities. The relationship sign with default risk is not clear cut because a high value of the ratio may indicate, on one hand, a larger positive contribution of extraordinary items and, on the other hand, that a larger part of gross profit has been made by extraordinary components (that typically do not reoccur in the future). The indicator is not structurally monotonic because both numerator and denominator can be positive or negative.

19. ***TaxesOnGross_Profit:*** taxes divided by gross of tax profit. The ratio indicates the tax burden the firm has to bear. This indicator does not have a meaningful relationship with default risk, as it strictly depends on the set tax rate. In any case, the ratio is not structurally monotonic because of the range of variability of the denominator that includes negative values.

20. ***IntangiblesOnTA:*** intangibles divided by total assets, shown as a percentage. Intangibles are usually considered the least liquid and the most volatile

type of assets. As such, the relationship with default risk is typically assumed as positive. The indicator is structurally monotonic.

21. *Trade_ReceivablesOnTA:* trade receivables divided by total assets, shown as a percentage. Trade receivables are seen as assets with a good degree of liquidity. Hence, it is commonly assumed that the probability of default decreases when the ratio increases. The indicator is structurally monotonic.

22. *InventoriesOnTA:* inventories divided by total assets, shown as a percentage. Setting a working hypothesis concerning its relation with default risk is not easy because inventories are more liquid than fixed assets but less liquid than other current assets. However, above all for SMEs, inventories often hide finished goods that have no market or sales that have not been regularly written in accounting books (hence, cash flows have been diverted from the firm). Therefore, a higher inventory per unit of assets is typically associated with a higher default risk. The indicator is structurally monotonic.

23. *EquityOnPermanent_Capital:* equity divided by permanent capital, shown as a percentage. The relationship with default risk is assumed to be negative because more equity is the safest form of capital from a creditors point of view. The indicator is structurally monotonic.

24. *Trade_PayablesOnTL:* trade payables divided by total liabilities, shown as a percentage. It indicates the structure of liabilities. Usually, a negative relationship with default risk is assumed considering that the firm whose suppliers are accepting longer payment terms has more resources to finance the business that are 'naturally' obtained through daily operations. The indicator is structurally monotonic.

25. *Debt_Equity_Ratio:* Debt divided by equity, shown as a percentage. Here, debt is used with an extensive meaning, as synonym of liabilities. It indicates the amount of liabilities per Euro of equity; as such, it has a positive relationship with default risk. In fact, a lower ratio implies both a safer financial structure and less volatile economic results due to lower financial leverage. As far as monotonicity is concerned, the peculiar decision to subtract firm's loans to shareholders from firm's equity leads the indicator to be structurally non-monotonic. Calculating the inverse would solve the problem.

26. *Current_Ratio:* current assets divided by current liabilities. It is an indicator of firm's liquidity: higher current assets per unit of current liabilities outline that more assets are going to become liquid in the short term than liabilities are going to become due in the short term. The relationship with default risk is negative. The indicator is structurally monotonic.

27. *Quick_Ratio:* current assets minus inventories, divided by current liabilities. It is a more specific indicator of firm's liquidity than the previous ratio, because it excludes inventories from current assets, that is to say, the type of current assets whose liquidation is more questionable. The relationship with probability of default is negative. The indicator is structurally monotonic.

28. *SalesOnVp:* sales divided by value of production, shown as a percentage. It indicates the share of value of production that is sold during the year, instead of being accumulated into inventories (when sales are lower than production) or taken from inventories (when sales are higher than production). Hence, the higher the ratio, the lower the probability of default. Only in very exceptional cases can the denominator be negative; therefore, we can assume that the ratio is monotonic.

29. *SalesMinusVcOnEBIT:* sales minus variable costs, divided by EBIT, shown as a percentage. It indicates the share of EBIT available to cover operating fixed costs. It is an indirect indicator of operating leverage risk. The relationship with default risk is negative, and the ratio is structurally monotonic.

30. *ROAminusIEonTL:* the difference between ROA and the ratio of interest expenses to total liabilities. The larger the difference, the lower the default risk, as it shows the firm's capability to invest financial resources with higher returns over their costs. The indicator is structurally monotonic. In general, it represents a buffer which protects the firm from possible negative spreads (due to increases in the cost of liabilities and/or decreases in ROA) that would create a loss for each Euro of liabilities invested in the company. Therefore, it is a cushion against unfavorable functioning of financial leverage. In fact, the following relationship holds:

$$\text{ROE} = [\text{ROA} + (\text{ROA} - \text{IEonTL}) \times \text{Debt/equity ratio}] \times K \times (1 - T)$$

where a number of indicators examined above are included (K is Extr_I_C_on-Current_Income and T is TaxesOnGross_Profit). The weak point of this formula in explaining financial leverage risk is that IEonTL distributes interest expenses over liabilities and not only over interest-bearing liabilities. A more sophisticated formula is the 'extended additive ROE formula' (De Laurentis, 1993):

$$\text{ROE} = [\text{ROA} + (\text{ROA} - \text{IEonFinancial\_Debt}) \times \text{D/E} + \text{ROA} \times \text{NIBLL}]$$
$$\times K \times (1 - T)$$

where two ratios that are not present in the dataset are included: D/E is the debt equity ratio correctly and strictly calculated having (interest-bearing) financial debt at numerator; NIBLL is the non-interest bearing liabilities leverage, calculated as non-interest bearing liabilities divided by equity. Note that K can be substituted by $(1- K')$, where K' is directly calculated as the ratio of 'extraordinary income and cost' to current income, that is to say, more similarly to the tax rate T. In this formula, financial leverage risk depends more explicitly on: (1) the volatility of ROA, basically the firm's business risk; (2) the volatility of C, which is the interest rate risk that the firm is facing (function of duration and interest rate renegotiation maturities of firm's borrowing, as well as of the volatility of market interest rates); this clarifies how interest rate risk can transform itself into default risk; it also clarifies that the bank has a trade-off between minimizing the former (extending medium and long term loans at a fixed interest rate) or increasing the latter; (3) the magnitude of the historical spread (ROA minus cost of debt), that protects the firm from unfavorable future evolutions of the return on assets and cost of debt; and (4) the correlation between ROA and cost of debt; it is structurally positive and strong for financial firms whose return on assets and cost of debt depend on the same market, the financial market; on the contrary, it is typically negative for non-financial firms because of the prevailing effect of risk-adjusted loan pricing policies of lenders that tends to increase the cost of borrowing when the firm's business is struggling. The brief examination just performed on financial leverage mechanics shows that the protection of a given amount of equity against unfavorable trends in a firm's economics depends on the four items listed above and cannot be appreciated by itself. The dataset of the case study includes only some of the relevant variables.

Analyses made up to now can be summarized in a summary table, such as Table 4.5. Operationally, it emerges that it is necessary:

1. to realize that indicators such as ROE and NIEonEBITDA are not structurally monotonic nor is there any simple algebraic transformation useful to solve the problem;

2. to consider calculating the reciprocal of IEonEBITDA and Debt_Equity_Ratio to obtain structural monotonicity;

3. to erase Extr_I_C_onCurrent_Income and TaxesOnGross_Profit because, even if they have a clear economic meaning, they do not have a meaningful relationship with probability of default;

4. to collect data concerning loans to shareholders in order to recalculate indicators using equity in a more traditional and useful way (that is to say, without subtracting firm's loans to shareholders). In addition, it would be useful to collect data about a more specific indicator of financial structure expressing a narrower concept of debt equity ratio where debt would only be financial debt and would not include all liabilities.

The first issue requires examining empirical monotonicity before, and then searching for a transformation that could solve the problems of non-monotonic indicators.

The second issue leads us to calculate two new variables. Open the following files:

W_CS_1_AnalysisSampleDataSet.sav,

W_AnalysisSampleSyntax_1.sps,

W_AnalysisSampleOutput_1.

From the SPSS-PASW main menu:

Transform, Compute variable, Variable name: for instance 'EBIT-DAonIE', *and in Numeric Expression: 1/IEonEBITDA*100, OK*

The new variable will be expressed in decimals (multiplying by 100 just offsets the percentage format of the original variable). In the 'data view', the new variable will be added to the right end of the dataset. If we sort the new variable in ascending order, we see many missing values. To check whether they were true zeros at the start (because EBITDA is zero) or whether they were missing values (because InterestExpenses are zero) that were transformed into zeros, add EBITDA and IE at the right end of the dataset as new temporary variables by:

> *Transform, Compute variable, Variable name:* for instance *'xEBITDA'*, *Numeric Expression: EBITDA, OK*

> *Transform, Compute variable, Variable name:* for instance 'xIE', *Numeric Expression: IE, OK*

The only observation while having EBITDA equal to zero and IE positive is BORCODE number 396. We can override the missing value in EBITDAonIE by typing zero directly into the cell; for all other observations missing values are due to IE equal to zero, so the ratio value should tend towards infinity, rather than being zero. For these cases, we can override missing data by setting very high values. Choose to set a value close to the 99th percentile of the distribution of the variable we now have. As the sample contains 1272 cases, we can simply sort the dataset in descending order according to the variable and choose the 13th highest value, that is: 131 (times). Now, sort in ascending order, type 131 in the first cell containing missing values and, then, copy and paste on all cells with missing values. We can eventually erase the two temporary variables, sort the dataset in descending order by BORCODE and:

> *File, Save as: W_CS_1_AnalysisSampleDataSet_2.sav, OK*

A similar transformation must be applied to 'Debt_Equity_Ratio', in order to obtain: 'EQUITYonLIABILITIES'. Also the transformed variable will be expressed in decimals:

> *Transform, Compute variable, Variable name:* 'EQUITYonLIABILI-TIES', and in *Numeric Expression: 1/* Debt_Equity_Ratio, *OK*

If we sort the new variable in ascending order, we see that there are seven missing values. To check whether they were true zeros at the start (because Liabilities are zero) or if they were missing (because Equity is zero) that were transformed into zeros, scroll the dataset up to Equity: it is zero for six of these variables and it has a positive value only for BORCODE 460. The six cases will be overridden by zeros, whereas the value of the variable for BORCODE 460, having a positive Equity and zero Liabilities, will be forced to the 13th higher value of the variable (3.55), as we did before.

The third issue which emerged from variables analyses simply leads to cancel the two variables from the dataset (Extr_I_C_onCurrent_Income and TaxesOnGross_Profit) or, at least, not to consider them in subsequent analyses.

The fourth issue requires us to try collecting new data. Assume we are able to acquire a new file containing two new variables:

1. *DebtEquityTr:* interest-bearing financial debt divided by equity (without subtracting loans to shareholders from total stockholders' equity), shown as a percentage. It corresponds to the D/E indicator used above in the 'extended additive ROE formula'. 'Tr' stands for 'traditional'. The ratio has a positive relation with default risk because the higher it is, the higher the volatility of ROE is and lower is the solidity of financial structure.

2. *ROETr:* net profit divided by total stockholders' equity, shown as a percentage. Therefore, the denominator is the 'traditional' equity, that is to say, without subtracting loans to shareholders.

In 'variable view', consider the first two empty rows and type the name of the two new variables. New data is in the Excel file named W_CS_1_Integra_ Database_Excel.xlsx. From it, we have to copy all variables' values and paste them in the 'data view'. Check that the SPSS-PASW file is sorted by BORCODE and that the entire range of cells to which to paste the data to is selected.
Save the W_CS_1_AnalysisSampleDataSet_2.sav file again: *File, Save.*

4.5.1.2   Categorical variables

Activities previously undertaken on financial ratios (understanding their economic meaning in order to set a working hypothesis on the sign of the expected relation with default risk and checking if they are structurally monotonic) have completely different content when considering categorical variables. Often, financial theory does not provide a clear and unidirectional working hypothesis, whereas empirical tests show a strong and statistically significant relation with default rates. This is the case for categorical variables such as the region in which an industrial firm is incorporated. In principle, there is no reason why different regions should express different default risk for companies, but often this is what is found empirically. Usually, there is a hidden relationship; for instance, the different location of business sectors across regions. Even in the absence of a specific recognized theory, sometimes it can be satisfactory to set the working hypothesis in line with long lasting documented empirical evidence the bank has (often explicitly incorporated into internal credit policies). For instance, some regions can be traditionally less risky than others: if data in the analysis sample confirm the historical behavior, we can assume that the variable 'region' has passed the test of economic meaningfulness. Usually, the same approach is taken also for business sectors, and their more or less granular aggregates, even if in principle it is easier to develop a theory explaining the propensity to go into default for such aggregates.
In addition, checking structural and empirical monotonicity of the variable has a completely different content when considering categorical variables. Structural monotonicity is meaningless. In the case of nominal variables, testing empirical

monotonicity is also meaningless because any single value of the variable should actually be considered as a different variable. This is apparent when you consider that levels of categorical variables (e.g., regions) are incorporated into multivariate analysis as a zero-one 'dummy variable' (is the company either operating in Region ABC or not? Is the company either operating in Region XYZ or not?). In the case of ordinal variables, the empirical monotonicity requirement comes into play again and is assessed using the same approach we will analyze later for scale variables.

4.5.2 Empirical assessment of working hypothesis

To analyze whether empirical data support working hypotheses for scale indicators, we can calculate, for each of the two groups of good and bad borrowers, descriptive statistics such as mean, trimmed mean, median, and so forth. For instance, as we said that we expect ROE to be inversely related with default risk, we should observe a higher ROE mean for the group of performing borrowers than for the group of defaulted ones. Such descriptive statistics can also suggest insights about the shape of the distribution of a given variable in each of the two groups and the possible existence of outliers impacting means.

Use the Explore routine of SPSS-PASW. It produces a number of statistics and graphs that will be useful later on. However, given the size of output files and the objectives of this paragraph, we will select statistics separately from graphs. From the SPSS-PASW main menu:

> *Analyze, Descriptive Statistics, Explore, Factor List: BadGood, Variables:* all financial ratios, *Display: Statistics,* in *Statistics* select: *Descriptives, Outliers, Percentiles* (Figure 4.7)

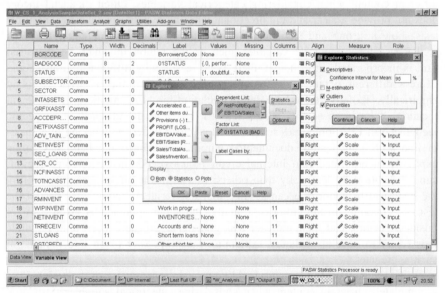

Figure 4.7   Descriptives using the Explore routine.

In the syntax window, the Explore routine is still named Examine, as in the initial releases of SPSS-PASW. Where percentiles are specified, add 2.5 as the first value and 97.5 as last one, in order to have a more detailed view of distributions' tails and to be able to focus on distributions' boundaries that exclude an overall 5% of observations with even more extreme values (Figure 4.8).

Figure 4.8   Changing the explore routine syntax.

Now, run Explore routine. In the output file, after '*Case Processing Summary*' table, we can read the results. The first table is '*Descriptives*'. It needs to be pivoted in order to better compare statistics for performing and defaulted borrowers:

> double click on *Table,* then select *Pivot, Pivoting Trays*: bring 01Status from the left bar to the upper bar, just below the existing icon, then close both dialog boxes to return to Descriptives table. Now, it has been reorganized for a better comparison of statistics for the two groups of observations. Let's save the output and syntax files as:
>
> *W_AnalysisSampleOutput_2stat, Save*
>
> *W_AnalysisSampleSyntax_2, Save*

Now, the comment will only consider three ratios (ROE, InterestExpenses/Liabilities, Inventories/TotalAssets). Readers using SPSS-PASW to develop analyses on all financial ratios are invited to fill in a table similar to Table 4.5, and then check results comparing them with Table 4.26 reported in Section 4.5.10. Table 4.6 reports the output of the Explore routine, pivoted as we did before, for the three considered ratios.

There are measures of:

(a) central tendency (mean, trimmed mean, median),

(b) variation (variance and standard deviation, minimum, maximum, range, interquartile range),

(c) shape (skewness and kurtosis),

(d) confidence interval for mean.

Mean is the arithmetic mean. Standard Error of the mean is a measure of how much the value of the mean may vary because of the sampling process (that is, from sample to sample taken from the same distribution). In fact, we are using a sample in order to infer conclusions on a larger population. Therefore, it would be useful to know how representative of the population mean a given sample mean is likely to be. Standard error is the standard deviation of means calculated on samples randomly taken from the population: a smaller standard error (relative to the sample mean) expresses greater similarity between sample mean and population mean. This statistic can also be used to roughly compare the observed mean to a hypothesized value: it is common to conclude that observed and hypothesized values are different if their difference divided by the standard error is less than -2 or greater than $+2$.

Confidence intervals are an alternative approach to assess how much a sample mean is representative of population mean. Interval boundaries represent the upper and lower bound (or limit) of a confidence interval where the population mean should fall, with a given probability (confidence level). Assuming a normal distribution, 68% of observations are included in the interval whose boundaries are mean plus/minus one standard deviation; 95% of observations are located between mean plus/minus 1.96 standard deviations. Therefore, if we assume, for instance, a 95% confidence level, we can assert that we are 95% confident that the population mean will be within the boundaries of the confidence interval calculated as sample mean plus/minus 1.96 standard deviations. The multiples of standard deviation associated with confidence levels are named 'critical values' of the distribution (1 or 1.96, and so on).

A 5% trimmed mean is the mean calculated on residual values, once the lowest 5% and highest 5% values have been excluded.

Median is the value above and below which half of the cases fall into; it is the 50th percentile.

A measure of dispersion around the mean is the (sample) variance. It is equal to the sum of squared deviations from the mean divided by the number of cases less one.

Standard deviation is the square root of the variance. In a normal distribution, 68% of the cases lie within an interval of plus and minus one standard deviation above and below the mean and 95% of the cases fall within 1.96 standard deviations. Higher standard deviations indicate a wider dispersion of values around the mean.

Table 4.6   Descriptives of three ratios.

			Statistic		Std. Error	
			01STATUS		01STATUS	
			performing	default	performing	default
NetProfit/	Mean		45.737	418.889	13.153	355.298
Equity	95% Confidence Interval for Mean	Lower Bound	19.932	−294.748		
		Upper Bound	71.542	1,132.526		
	5% Trimmed Mean		19.046	24.639		
	Median		10.000	0.487		
	Variance		211,241.235	6,438,063.07		
	Std. Deviation		459.610	2,537.334		
	Minimum		−5, 500.000	−525.000		
	Maximum		8,300.000	17,930.000		
	Range		13,800.000	18,455.000		
	Interquartile Range		42.459	30.635		
	Skewness		5.824	6.854	0.070	0.333
	Kurtosis		140.094	47.989	0.140	0.656
Interest	Mean		3.820	6.487	.073	2.039
Expenses/	95% Confidence Interval for Mean	Lower Bound	3.676	2.391		
Liabilities		Upper Bound	3.964	10.583		
	5% Trimmed Mean		3.660	4.506		
	Median		3.480	4.524		
	Variance		6.592	212.076		
	Std. Deviation		2.567	14.563		
	Minimum		.000	0.000		
	Maximum		23.950	107.407		
	Range		23.950	107.407		
	Interquartile Range		3.259	3.051		
	Skewness		1.297	6.918	0.070	0.333
	Kurtosis		4.507	48.826	0.140	0.656
Inventories/	Mean		20.852	26.912	0.665	3.337
Total	95% Confidence Interval for Mean	Lower Bound	19.546	20.210		
Assets		Upper Bound	22.157	33.614		

(continued)

Table 4.6   *(continued)*

	Descriptives			
	Statistic		Std. Error	
	01STATUS		01STATUS	
	performing	default	performing	default
5% Trimmed Mean	18.336	24.933		
Median	12.832	22.537		
Variance	540.714	567.823		
Std. Deviation	23.253	23.829		
Minimum	0.000	0.000		
Maximum	101.529	95.888		
Range	101.529	95.888		
Interquartile Range	28.792	30.823		
Skewness	1.425	1.098	0.070	0.333
Kurtosis	1.473	0.839	0.140	0.656

Minimum is the smallest (lowest) value.

Maximum is the largest (highest) value.

Range is the difference between maximum and minimum values.

Interquartile range is equal to the difference between the third quartile and first quartile. Hence, it includes 50% of the data and represents the spread in the middle 50% of the data.

Skewness is a measure of asymmetry of a distribution. The normal distribution is symmetric and has a skewness value of 0. A distribution with a significant positive skewness has a long right tail. A distribution with a significant negative skewness has a long left tail. As a guideline, a skewness value more than twice its standard error indicates a departure from symmetry.

Kurtosis is a measure of the extent to which observations cluster around a central point. Positive kurtosis indicates that the observations cluster more (and have longer tails) than those in the normal distribution, and negative kurtosis indicates that the observations cluster less and have shorter tails. In many standard formulas, for a normal distribution, the value of the kurtosis statistic is three, but SPSS-PASW uses a different formula to calculate kurtosis, so here, a normal distribution has the value of the kurtosis statistic equal to zero.

Standard errors of skewness and kurtosis depend on sample size and not on the specific ratio under examination. This is why in SPSS-PASW output standard errors of skewness and kurtosis are identical for all variables. These measures can offer more precise information on the probability that these two measures of shape indicate non-normal distributions. Consider skewness (the same applies to kurtosis): it is common to conclude that the distribution is not normal if the skewness statistic divided by its standard error is less than -2 or greater than $+2$. In fact, we can calculate standardized 'z' values that can be compared with critical values of a standardized normal distribution in order to answer if the statistical hypothesis is significant. Let's calculate:

z = (sample statistic − hypothesized value)/standard error of the statistic.

If we assume the typical confidence level of 95%, critical values are ± 1.96 (often rounded to ± 2); these indicate confidence interval boundaries. If z is within the confidence interval, the statistical hypothesis is not rejected. In fact, in this case, the following equivalent conclusions can be drawn:

(a) we are 95% certain that the sample statistics are genuinely close to the hypothesized value,

(b) there is only a 5% probability that the result occurred by chance.

In the case of skewness (S), in order to test if it is zero, we calculate $z = (S - 0)/SE_S$, where SE_S is the standard error of skewness. If 'z' (that simply is S/SE_S) is less than -2 or greater than $+2$ it means that it falls outside the confidence interval, that is to say, in the rejection areas of statistical hypothesis: in this case we should conclude that the distribution is not normal because of skewness. However, for large samples (say over one thousand observations) SE_S tends to become smaller and, as a consequence, z values are typically outside critical values even for small deviations from normality. This is why, for large samples, it is better to visually judge the distribution (by box plots or histograms and the like) rather than using skewness statistics and associated z values.

Now, let's comment on the results for the three financial ratios used as exemplificative cases.

Mean, trimmed mean, and median are all aligned with working hypothesis except in the case of ROE, for which mean and trimmed mean are conflicting with working hypothesis (defaulted borrowers have higher values than performing borrowers). The large difference between these statistics outlines the existence of outliers. This phenomenon is particularly strong for ROE: the average ROE for defaulted borrowers is not only higher than for performing ones but it also has an unrealistically high value (419%).

Confidence intervals of the means of good and bad borrowers of the three ratios are largely overlapped. Standard errors of means are very high.

Furthermore, measures of variations indicate a large dispersion of distribution values. The shape of distributions is far from normal. Skewness and Kurtosis are always positive. Only for inventories divided by total assets – in the case of bad borrowers – a normal shape cannot be excluded at a 95% confidence level.

Percentiles distribution (Table 4.7) confirms the wide range of variability of distributions, particularly of ROE: just outside the interquartile range, ratios' values increase exponentially.

Table 4.7 Percentiles.

	01STATUS	Percentiles								
		2.5	5	10	25	50	75	90	95	97.5
Weighted Average (Definition 1)										
NetProfit/Equity	performing	−310.692	−143.130	−57.330	0.000	10.000	42.459	109.080	216.373	595.083
	default	−460.500	−193.421	−84.789	−12.453	0.487	18.182	136.182	1,536.915	13,399.936
Interest Expenses/ Liabilities	performing	0.000	0.308	0.835	1.935	3.480	5.195	7.047	8.378	9.623
	default	0.164	1.194	2.085	2.987	4.524	6.038	7.276	9.735	78.148
Inventories/ TotalAssets	performing	0.000	0.000	0.000	2.566	12.832	31.358	55.297	75.167	86.520
	default	0.000	0.064	0.583	8.490	22.537	39.313	64.823	83.422	93.495
Tukey's Hinges										
NetProfit/Equity	performing				0.000	10.000	42.308			
	default				−9.926	0.487	17.694			
Interest Expenses/ Liabilities	performing				1.935	3.480	5.194			
	default				3.177	4.524	5.871			
Inventories/Total Assets	performing				2.601	12.832	31.348			
	default				9.225	22.537	38.216			

Overall, the majority of the observed ratios do not have problems of coherence with working hypotheses when trimmed means and medians are considered, whereas the opposite is true for means.

Consider now the two categorical values in the dataset: subsector and sector. We have already observed that subsector levels are poorly populated. Therefore, let's focus on sector and let's verify whether it offers useful information on default risk. Of course, the analytical approach is different from that used for scale variables. Here, the frequency of default for each level of the categorical variable must be calculated and compared. From the SPSS-PASW main menu:

Analyze, Descriptive Statistics, Crosstabs, Row: 01Status, Column: Sector, *Statistics:* none, *Continue, Cells: Observed, Percentage: Row, Total, Continue, Paste*

The script obtained is:

```
CROSSTABS
 /TABLES=Sector BY BadGood
 /FORMAT=AVALUE TABLES
 /CELLS=COUNT ROW TOTAL
 /COUNT ROUND CELL.
```

Run the new syntax and restructure the cross-tabulation in the output viewer by using *Pivot* and *Pivoting trays* commands in order to obtain Table 4.8. It is evident that there is variability in default rates (sixth column from the left); as we have already said, for categorical variables the concept we can assimilate to a working hypothesis is the alignment of sample results with historical data and beliefs emerged in the bank and in external studies in the past. In any case, many sector codes are poorly populated.

In the next sections, statistical tests will be used to check the statistical significance of: means and medians differences in bad and good subsamples, normality of distributions and association between categorical variables levels and default rates.

4.5.3 Normality and homogeneity of variance

Some multivariate techniques used to estimate statistical based rating systems require explanatory variables to show normal distributions and homogeneity of variance in the two subgroups. This is the case for linear discriminant analysis. Apart from graphical analysis, the normality of bad and good subsamples distributions can be assessed using Kolmogorov–Smirnov and Saphiro–Wilk statistical tests. They compare values of the financial ratio under examination with values that a normal distribution which has the same mean and standard deviation would present. The null hypothesis for both tests is that there is no difference between the observed sample distribution and a normal distribution: the hope is to find a larger

Table 4.8   Default rates by sectors.

		Count			% within SectorCode			% of Total		
		01STATUS			01STATUS			01STATUS		
		performing	default	Total	performing	default	Total	performing	default	Total
Sector	258	1	0	1	100.0	0.0	100.0	0.1	0.0	0.1
Code	268	2	0	2	100.0	0.0	100.0	0.2	0.0	0.2
	273	1	0	1	100.0	0.0	100.0	0.1	0.0	0.1
	280	1	0	1	100.0	0.0	100.0	0.1	0.0	0.1
	294	1	0	1	100.0	0.0	100.0	0.1	0.0	0.1
	430	712	33	745	95.6	4.4	100.0	56.0	2.6	58.6
	431	1	0	1	100.0	0.0	100.0	0.1	0.0	0.1
	450	4	0	4	100.0	0.0	100.0	0.3	0.0	0.3
	470	1	0	1	100.0	0.0	100.0	0.1	0.0	0.1
	473	2	0	2	100.0	0.0	100.0	0.2	0.0	0.2
	480	13	0	13	100.0	0.0	100.0	1.0	0.0	1.0
	481	3	0	3	100.0	0.0	100.0	0.2	0.0	0.2
	482	236	7	243	97.1	2.9	100.0	18.6	0.6	19.1
	490	17	1	18	94.4	5.6	100.0	1.3	0.1	1.4
	491	1	0	1	100.0	0.0	100.0	0.1	0.0	0.1
	492	114	4	118	96.6	3.4	100.0	9.0	0.3	9.3
	600	4	0	4	100.0	0.0	100.0	0.3	0.0	0.3
	614	71	5	76	93.4	6.6	100.0	5.6	0.4	6.0
	615	35	1	36	97.2	2.8	100.0	2.8	0.1	2.8
Total		1,220	51	1,271	96.0	4.0	100.0	96.0	4.0	100.0

p-value than the set significance level (usually 5%) in order to avoid rejection of the null hypothesis. To summarize:

(a) Ho: no difference with a normal distribution

(b) desired p-value > 0.05.

Homogeneity of variance between the subsamples of good and bad observations can be assessed by Levene's test. Its null hypothesis is that variances are homogenous: the hope is to find a larger p-value than the set significance level (5%) in order to avoid rejection of the null hypothesis. To summarize:

(c) Ho: no difference of variances of distributions

(d) desired p-value > 0.05.

In SPSS-PASW, the same Explore routine we used before produces statistical tests (as well as graphs that we are going to examine in the next section). We could have selected the whole set of outputs at once, but for the sake of clarity we prefer to select them step by step. Close the W_*AnalysisSampleOutput_2stat* file in order to produce a new output file for the new analysis. From the SPSS-PASW main menu (Figure 4.9):

> *Analyze, Descriptive Statistics, Explore, Factor List: BadGood, Variables:* all ratios, *Display: Plots,* in *Plots* choose: *Factor levels together, Normality plots with test,* In *Spread vs Level with tests* choose *Untransformed, Paste*

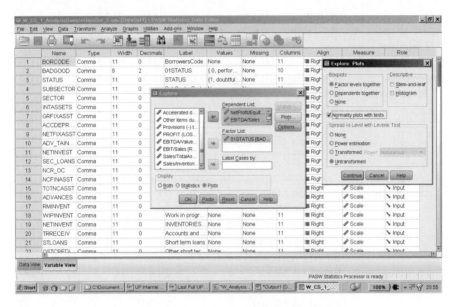

Figure 4.9   Plots and tests in the explore routine.

Do not select *stem & leaf, histogram* or other analyses; they are alternative representations of distribution shape that consume more space and are less clear. Select the new syntax and *Run* it. Save the output and syntax files as:

> *W_AnalysisSampleOutput_2gratest, Save*
>
> *W_AnalysisSampleSyntax_2, Save*

As usual, we will only comment on the three ratios we have adopted as examples. Table 4.9 indicates that we have to reject the hypothesis of normal distributions for all examined ratios because the observed level of significance is always less than 0.05.

In consideration to homogeneity of variance, only Inventories/TotalAssets meets the requirement.

If all ratios in the dataset are considered, tests show that a large majority of them do not meet the normality requirement whereas about two-thirds meet the homogeneity of variance requirement.

Table 4.9   Normality and homogeneity of variance tests.

		Tests of normality					
01STATUS		Kolmogorov–Smirnov[a]			Shapiro–Wilk		
		Statistic	df	Sig.	Statistic	df	Sig.
NetProfit/Equity	performing	0.348	1221.000	0.000	0.267	1221	0.000
	default	0.466	51.000	0.000	0.187	51	0.000
InterestExpenses/ Liabilities	performing	0.068	1221.000	0.000	0.930	1221	0.000
	default	0.398	51.000	0.000	0.229	51	0.000
Inventories/ TotalAssets	performing	0.185	1221.000	0.000	0.826	1221	0.000
	default	0.129	51.000	0.033	0.899	51	0.000

	Test of homogeneity of variance				
		Levene statistic	df1	df2	Sig.
NetProfit/Equity	Based on Mean	54.077	1	1270	0.000
	Based on Median	14.686	1	1270	0.000
	Based on Median and with adjusted df	14.686	1	152	0.000
	Based on trimmed mean	15.378	1	1270	0.000
InterestExpenses/ Liabilities	Based on Mean	28.473	1	1270	0.000
	Based on Median	12.293	1	1270	0.000
	Based on Median and with adjusted df	12.293	1	91	0.001
	Based on trimmed mean	12.337	1	1270	0.000
Inventories/ TotalAssets	Based on Mean	0.113	1	1270	0.736
	Based on Median	0.244	1	1270	0.622
	Based on Median and with adjusted df	0.244	1	1268	0.622
	Based on trimmed mean	0.173	1	1270	0.677

[a]Lilliefors Significance Correction

4.5.4 Graphical analysis

The Explore routine produces a number of graphs together with tests and descriptive statistics we have already examined. Many aspects concerning distribution shape and relative positions of central tendency measures, which we have already observed through descriptive and statistical tests, can also be analyzed through graphs. This approach is often more straightforward but even more subjective. Graphical analysis is particularly useful when:

(a) trying to spot outliers. In fact, we do not yet know their number and their distribution through borrowers (in terms of BORCODE) and ratios.

(b) The sample's size is large (for example, beyond one thousand observations). In fact, in such cases, tests which have been examined above and used to check distributions normality and homogeneity of variance tend to produce very high statistics and, as a consequence, very low p-values. This implies rejection of null hypotheses: in other words, they detect statistically significant but unimportant deviations from normality or equality of variance. Levene's test can be substituted using the Variance Ratio (the variance of the group with the higher value of variance divided by the variance of the group with the lower value): if it is less than two, we can conclude that there is homogeneity of variance. On the other hand, the normality test can be replaced with or put beside graphical examination.

There are many types of graphs; some of which include: Box plot, Q–Q plot, P–P plot, Histograms, Stem and leaf. Let us use the first two.

Box plots, sometimes called box-and-whiskers plots, shows the median, quartiles, outliers, and extreme values for scale variables. The interquartile range (IQR, the difference between the 75th and 25th percentiles) corresponds to the length of the box. Outliers are values between $1.5 \times IQR$ and $3 \times IQR$ from the box's boundaries, and are indicated by a circle and a label representing the progressive number of the observation in the dataset. Extremes (defined here as values more than $3 \times IQR$ from the end of a box) are indicated by an asterisk and a label representing the progressive number of the observation in the dataset again. The extreme point of the whisker indicates the value of the last observation before the first outlier.

We can verify the following by using box plots:

(a) distribution symmetry around median,

(b) outlier and extreme values,

(c) extension of groups' distributions overlapping. The lower the overlapping, the higher the discriminatory power of the ratio (further on, it will be quantitatively expressed by F-ratio and AuROC measures).

Examine the result in the output viewer. The box plot for ROE is extremely compressed because of outliers and extremes (Figure 4.10, left side). SPSS-PASW allows users to zoom in order to have a clearer box plot, such as the one which can be seen on the right side of Figure 4.10. It is better to manipulate a copy of

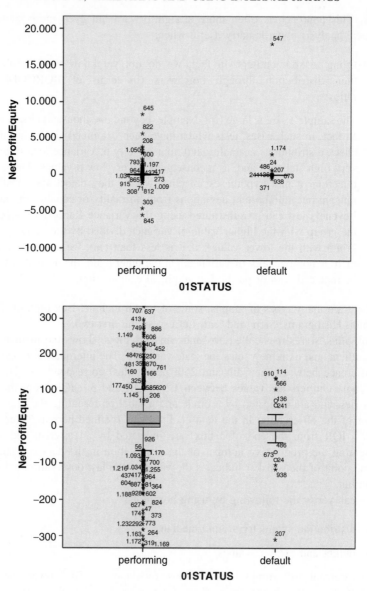

Figure 4.10   A box plot.

the original output; therefore, *Copy, Paste After* and, once we have the copy, select it and:

> right click on boxplot, *Edit Content, In separate window, Edit, Rescale chart,* and using icon *Rescale Chart* to select the area to enlarge, then *Close* (Figure 4.11).

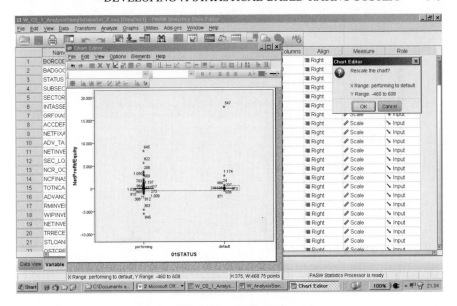

Figure 4.11   Rescaling the box plot.

An alternatively and sometimes more effective routine is:

right click on Boxplot, *Edit Content, In separate window, Edit, Select Y axis, Scale* and in *Range,* set *minimum* and *maximum* as large but reasonable values for the ratio.

The number of outliers and extremes shown in Figure 4.10 on both tails of the distribution of performing borrowers' ROE is impressive. The two distributions completely overlap, even if the interquartile range of performing borrowers' distribution appears correctly positioned above defaulted borrowers' distribution, leaving some hope for the discriminatory power of the variable. Also, with the new zoom level, the box plot symmetry remains difficult to be interpreted; perhaps a Q–Q plot can be more useful.

The indicator of average cost of liabilities has a unique extreme value in the bad borrowers' subsample and interquartile ranges are correctly positioned even if the overlapping area appears to be large (Figure 4.12). This last observation also applies to Inventories/TotalAssets.

A Q–Q plot graphically compares empirical distribution quantiles with quantiles of a theoretical distribution (a normal distribution for us). If the selected variable matches the test distribution, the points cluster around a straight line (Figure 4.13).

When adopting a good level of benevolence, the Inventories/TotalAssets distribution for defaulted borrowers can be considered almost normal. If we evaluate all indicators in the long list via Q–Q plots, a dozen distributions, sometimes for good borrowers and more often for bad borrowers, can be considered normal.

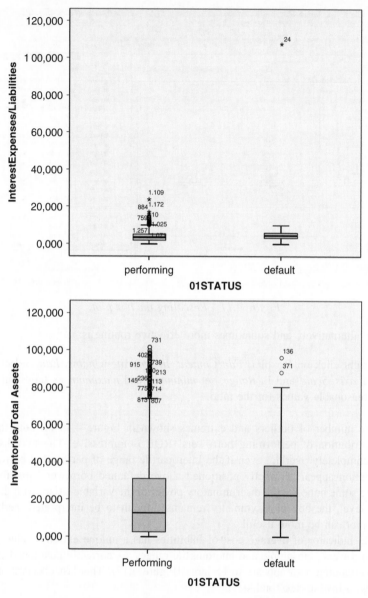

Figure 4.12   Rescaled box plot for cost of liabilities.

In conclusion, graphical analysis slightly increases the number of normal distributions identified by statistical tests. Box plot analysis has clearly outlined the existence of many outliers and extreme values, and many ratios with extensive overlapping areas of the distributions for good and bad cases. In the next setion, this last aspect is examined more in depth.

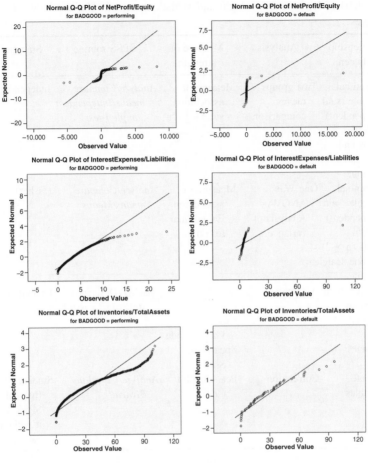

Figure 4.13   A Q–Q plot.

4.5.5   Discriminant power

For an indicator to be used in a statistical based rating system, it has to exhibit an acceptable level of discriminatory power between good and bad borrowers. Up to now, we have used neither the measures of strength of relationship between explanatory variables and the 01Status variable, nor the related statistical tests. Table 4.10 presents a summary table of measures of strength and tests applicable to different combinations of scale and categorical variables.

An independent t-test is used to compare means of a scale variable across groups identified by a categorical variable when it has only two levels (so there are two mutually exclusive groups). ANOVA (analysis of variance) is used when the categorical variable identifies two or more groups. Later, we will also consider ROC curve and CAP (and their related quantitative measures AuROC and Gini ratio)

Table 4.10   Measures of strength and statistical test on relations between variables.

Relationship between	Analysis	Measure of strength	SPSS routine	Statistical test
One nominal variable (said independent) with two levels and one scale variable	Two groups means comparison	Measure of association. similar to 'r' $r = \sqrt{\frac{t^2}{t^2+df}}$	*Analyze/Compare means/Independent sample t-test*	Independent t-test
One nominal variable (said independent) with 'n' levels (n $\geq$ 2) and one scale variable	One Way ANOVA – Analysis of variance	Measure of association Eta (similar to 'r')	*Analyze/Compare means/Means*	(Fisher) F-test
Two categorical variables nominal ordinal	Association	 Cramer's V. Phi. Lambda Spearman. Somers' d	*Analyze/Descriptive statistics/Crosstabs*	Chi-squared
Two scale variables	Correlation	Pearson Rho ('r')	*Analyze/Correlate/ Bivariate*	Student t-test (two tails). SPSS produces it without naming it

for the same task, as they are widely used in credit risk management to analyze the relationship of a nominal variable (such as 01Status) both with scale variables (such as financial ratios) and with ordinal variables (such as rating classes).

Association analysis examines the proportion of good and bad cases (default rates) across different levels of a nominal variable (such as sector or subsector).

Correlation analysis checks if pairs of ratios in the long list give the same signals and, consequently, if one of the two ratios is redundant in order to be considered in the short list.

Start with the Independent Sample t-test.

Analyze, Compare means, Independent sample T test, Test variables: all financial ratios, *Grouping variable: 01Status, Define groups:* 0 e 1, *Paste*

Once syntax has been created, focus only on the three variables we are using as an example in the book. Table 4.11 reports descriptives concerning variables that we have already discussed.

Table 4.11   Descriptives preceding the t-test.

Group Statistics						
01STATUS			N	Mean	Std. Deviation	Std. Error Mean
NetProfit/Equity	dimension1	performing	1,221	45.737	459.610	13.153
		default	51	418.889	2,537.334	355.298
InterestExpenses/	dimension1	performing	1,221	3.820	2.567	0.073
Liabilities		default	51	6.487	14.563	2.039
Inventories/	dimension1	performing	1,221	20.852	23.253	0.665
TotalAssets		default	51	26.912	23.829	3.337

Table 4.12 provides indications about the statistical significance of means difference for three ratios across the bad and good groups of observations. Hence, it permits verification of ratios having statistically significant discriminatory power.

First of all, it is necessary to decide which row to consider for each ratio. The decision depends on Levene's test, which concerns homogeneity of variance between groups. Remember that the null hypothesis is that there is homogeneity. Therefore, if the p-value (Sig. in the third column of the table) is less than or equal to the set level of significance (usually 0.05, known as the α-value, complement to one of confidence level), the null hypothesis is rejected and the second row must be considered. This is the case for ROE and cost of liabilities. If the p-value is more than 0.05, the first row is relevant (this is the case for Inventories/TotalAssets). Once the relevant rows have been selected for each ratio, we can see that the t-test p-value (sixth column in Table 4.12) is more than 0.05 for all the three ratios which have been considered. Now, t-test functioning is as follows:

(a) Ho: no difference between means

(b) desired P-value $\leq$ 0.05.

Therefore, as t-test p-values in Table 4.12 are always more than the set level of significance, it means that the t-statistic (fourth column in the table) falls in the region of non-rejection of the null hypothesis. Therefore, as we cannot reject the null hypothesis, we conclude that the observed means difference between groups could have been obtained by chance.

If we consider the entire long list of indicators, we realize that only three of them have means difference statistically significant at the 0.05 level: TradeReceivables/TotalAssets, TradePayables/TotalLiabilities, EBITDAonIE.

If the t-test is statistically significant, then it becomes useful to measure the strength of the relationship. SPSS-PASW does not produce this measure. However,

Table 4.12 Independent samples test.

| | | Independent samples test | | | | | | | | |
| | | Levene's test for equality of variances | | t-test for equality of means | | | | | 95% Confidence Interval of the Difference | |
		F	Sig.	t	df	Sig. (2-tailed)	Mean Difference	Std. Error Difference	Lower	Upper
NetProfit/Equity	Equal variances assumed	54.077	.000	−3.865	1,270	.000	−373.152	96.554	−562.574	−183.729
	Equal variances not assumed			−1.050	50	.299	−373.152	355.541	−1,087.229	340.925
Interest Expenses/Liabilities	Equal variances assumed	28.473	.000	−4.870	1,270	.000	−2.667	.548	−3.741	−1.592
	Equal variances not assumed			−1.307	50	.197	−2.667	2.041	−6.765	1.431
Inventories/TotalAssets	Equal variances assumed	.113	.736	−1.822	1,270	.069	−6.060	3.327	−12.587	.466
	Equal variances not assumed			−1.781	54	.080	−6.060	3.402	−12.882	.761

we can calculate it by the following formula, which uses the t-statistic and degree of freedom reported in Table 4.12:

$$r = \sqrt{\frac{t^2}{t^2 + df}}$$

The result is a sort of correlation coefficient 'r'. Usually, an absolute value of 'r' higher than 50% indicates a strong relationship, a value close to 30% expresses an average correlation. TradeReceivables/TotalAssets and EBITDAonIE have an average strength of the relationship of mean with bad and good groups, whereas TradePayables/TotalLiabilities has a weak relationship.

If ANOVA is used, the F-test verifies the null hypothesis of no difference among groups' means. Therefore, the aim is to have p-values less than 5% in order to reject null hypothesis. To summarize:

(a) Ho: no difference among groups means of a given scale explanatory variable

(b) desired p-value < 0.05.

Statistic value (F-ratio) is a good indicator of the strength of discriminatory power of a financial ratio: the higher its value, the stronger the relationship and the discriminatory power. In fact, the F-ratio is equal to the mean squares of the model divided by the mean squares of residuals (MS_M/MS_R). Mean squares are the sum of squares divided by the degrees of freedom. In particular, MS_M is the 'model mean square' (indicative of 'among groups' variance'). The degrees of freedom are the number of groups (k) minus one; we have only two groups (good and bad borrowers, identified by the nominal variable 01Status). The model sum of squares (SS_M) is the sum of squared differences between each group mean ($\bar{x}_k$) and the grand mean ($\bar{x}_{grand}$), multiplied by the number of observations in each group (n_k): $SS_M = \sum n_k(\bar{x}_k - \bar{x}_{grand})^2$. MS_M indicates the variance explained by the model (systematic variation), where the concept of 'model' is simply the sample subdivision in bad and good groups. In fact, the subdivision can 'explain' a part of the overall variation of values of a given financial ratio; MS_R is the 'residual mean square' (indicative of 'within groups' variance'). The degrees of freedom are the total sample size (N) minus the number of groups (k). The residual sum of squares (SS_R) is the sum of squared differences between each data (x_{ik}) and group mean: $SS_R = \sum(x_{ik} - \bar{x}_k)^2$. MS_R indicates the variance of values of a given financial ratio for each borrower around the group's mean. It is the residual variance, the part that cannot be explained by the 'model' as it is due to unsystematic factors.

The higher the F-ratio, the more the nominal variable under examination explains the 'grand variance' (the variance of all observations); that is to say, the groups' means explain a large part of overall variance, whereas the variance of individual observations around groups' means is low. It signifies that the means of two groups are significantly different and distinctly represent respective groups. The F-ratio is, therefore, a measure of the strength of a relationship but, as opposed to Eta, it can be used to compare financial ratios only by using an identical sample: in this case study, this condition is met within the analysis sample as we do not have any missing value.

There are various alternatives to obtain these measures by SPSS-PASW. The most straightforward and complete alternative is:

> *Analyze, Compare Means, Means, Dependent list:* all financial ratios, *Independent list:* 01Status, *Options: ANOVA Table Eta, Statistics: Mean, Standard deviation, Continue, Paste*

It is not useful to select *Count also,* because we already know that for all ratios we have 1221 performing borrowers and 51 defaulted borrowers, for a total amount of 1272 observations. Run the syntax and then reorganize the table with means and standard deviations:

> double click using the left button of the mouse, *Transpose rows and columns, Pivot, Pivoting trays,* move statistics on top bar above 01Status, *Close.* Save as:

> *W_AnalysisSampleOutput_3*

> *W_AnalysisSampleSyntax_3*

ANOVA is reported in Table 4.13 and measures of association in Table 4.14.

Table 4.13   ANOVA.

ANOVA Table							
			Sum of Squares	df	Mean Square	F	Sig.
NetProfit/ Equity* 01STATUS	Between Groups	(Combined)	6,816,633.235	1	6,816,633.235	14.936	0.000
	Within Groups		579,617,460.129	1270	456,391.701		
	Total		586,434,093.364	1271			
Interest Expenses/ Liabilities* 01STATUS	Between Groups	(Combined)	348.176	1	348.176	23.715	0.000
	Within Groups		18,645.629	1270	14.682		
	Total		18,993.805	1271			
Inventories/ TotalAssets* 01STATUS	Between Groups	(Combined)	1,798.058	1	1,798.058	3.319	0.069
	Within Groups		688,062.642	1270	541.782		
	Total		689,860.699	1271			

Note that the mean differences for ROE and cost of liabilities appear to be statistically significant in this analysis. The problem is that ANOVA assumes (among

Table 4.14   Measures of association.

Measures of association		
	Eta	Eta squared
NetProfit/Equity * 01STATUS	0.108	0.012
InterestExpenses/Liabilities * 01STATUS	0.135	0.018
Inventories/TotalAssets * 01STATUS	0.051	0.003

other assumptions) that there is homogeneity of variance in the two groups and, therefore, it tends to be misleading when this assumption is violated. Table 4.14 reports Eta and Eta-squared measures, which are similar to Pearson's 'r' coefficient of correlation. An Eta around 10% indicates a weak association. Inventories/TotalAssets has an even lower Eta.

Now, consider categorical explanatory variables such as 'Sector'. Beyond the observation of default rates for different levels of the categorical variable, it is useful to develop measures of association and their relative statistical tests. They can be obtained by:

Analyze, Descriptive Statistics, Crosstabs, Row: 01Status, *Column:* Sector, *Statistics: Chi-squared, Phi and Cramer's V, Continue, Cells: Observed, Percentage: Row, Total, Continue, Paste*

The (Pearson) Chi-square test essentially tests whether two categorical variables forming a contingency table are associated. If the observed level of significance (p-value) indicated in the right end column of Table 4.15 is below 0.05, it is inferred that a value of the test statistic is so big that it is unlikely to have happened by chance: the strength of the relationship is significant. This is not the case in our data. In addition, the Chi-square test requires that a low number (say 20% in larger contingency tables) of cells have an expected frequency below five: we have more than 76% of them.

Table 4.15   Chi-square tests.

Chi-Square Tests			
	Value	df	Asymp. Sig. (2-sided)
Pearson Chi-Square	4.280[a]	18	1.000
Likelihood Ratio	5.566	18	0.998
Linear-by-Linear Association	0.035	1	0.852
N of Valid Cases	1271		

[a] 29 cells (76.3%) have expected count less than 5. The minimum expected count is 0.04.

In any case, the Chi-square test does not suggest anything about how strong the association may be. Phi and Cramer's V are measure of strength of association and their significance says how probable it is that the measure has happened by chance (Table 4.16). The former is used for 2×2 contingency matrices; the latter occurs when variables have more than two categories. In our data we have a 100% probability to have obtained the measure of association by chance.

Table 4.16   Phi and Cramer's V measures.

Symmetric measures			
		Value	Approx. Sig.
Nominal by Nominal	Phi	0.058	1.000
	Cramer's V	0.058	1.000
N of Valid Cases		1271	

The discriminatory power of a scale variable (as well as of ordinal variables and, as we shall see, of scoring models) can be quantified using ROC curves and AuROC measures. Lorenz curves (also called CAP curves, cumulative accuracy profile) and Gini coefficients (also called AR, accuracy ratios) are similar but not identical tools.

A ROC curve offers a graphical representation of a financial ratio values capability to correctly distinguish bad borrowers from good ones. First of all, borrowers must be ranked according to their financial ratio values, from the riskiest to the safest. Then a graph is drawn indicating, for each rank:

- on Y-axis (Sensitivity) the cumulated frequency of defaults having a rank equal or lower the one considered;

- on the X-axis (1−Specificity) the cumulated frequency of performing borrowers having a rank equal or lower to the one considered.

Table 4.17 shows data for a given financial indicator: in the first column are indicators' values ordered from the lowest (assumed to indicate the riskiest level) to the highest; in the second column is the cumulated percentage of defaulted borrowers having a ranked value of the indicator equal or lower than that indicated in the first column; the third column indicates the cumulated percentage of performing borrowers which have a ranked value of the indicator equal or lower than that indicated in the first column. The second and third columns are, respectively, the Y-axis and the X-axis of the ROC curve in Figure 4.14.

Sensitivity is the ability to detect bad cases (or the probability of correctly identifying bad cases). Specificity is the ability to detect good cases (or the probability of correctly identifying good cases); therefore, 1−specificity means that you read specificity from right to left, and when reading from the left to the right you have an indication of 'the lack of specificity'.

A financial ratio with a great discriminatory power would rapidly cumulate bad borrowers on the Y-axis, while cumulating few performing borrowers on the

Table 4.17   Values on ROC curve axes.

	Sensitivity	1 – Specificity
−20,3900	,000	,013
−19,8230	,020	,013
−19,7595	,020	,014
−19,4825	,020	,014
−19,1040	,020	,015
−18,0010	,020	,016
−16,6025	,020	,017
−15,9970	,039	,017
−15,5215	,039	,018
−15,0775	,039	,019
−14,4065	,039	,019
−13,5220	,039	,020
−12,7065	,039	,021
−11,8805	,039	,022
. . .	. . .	. . .

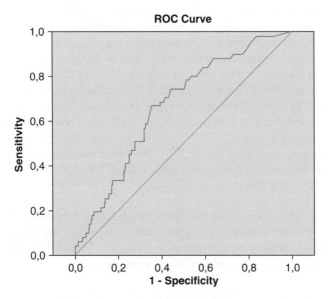

Figure 4.14   A ROC curve.

X-axis. A ROC curve would steadily increase almost vertically, with a limited shift to the right end of the graph, until almost all bad borrowers would have been cumulated (that is to say, detected).

As an example, suppose we are testing the capability of a chemical component (it represents a financial ratio in our case study) to selectively kill weeds (bad borrowers) without killing grass (good borrowers). We can empirically test the ratio of weed that is killed (sensitivity) and the ratio of grass that is killed (lack of specificity) for each concentration level of the chemical component.

If a financial ratio is able to perfectly separate non-defaulting and defaulting borrowers, the curve reaches 100% of defaults on the Y-axis while having considered none of the performing borrowers on the X-axis. On the other hand, if the ratio has no discriminatory ability at all, the ROC curve falls along the diagonal because the same percentages of good and bad borrowers are identified for each value of the ratio. Actual financial ratios stay in the middle: the more a curve is vertical, the better the ratio is.

A synthetic measure of discriminatory power is AuROC, which represents the area under the ROC curve. If we find a financial ratio with an AuROC of 100% (that is one, like saying 1×1, that is the entire box, because the ROC curve has increased vertically up to 100% of bad borrowers before considering any good borrower), it means that the financial indicator is a perfect explanatory variable. If the ROC curve is moving alongside the diagonal, the AuROC is 50%. It indicates that the variable is not adding any valuable information to distinguish between bad and good borrowers, and using it as a discriminant tool would be the same as randomly assigning borrowers to the two groups. In other words, once borrowers are ranked on the basis of that financial indicator, we would pick up good and bad borrowers for any given rank with the same proportion by which they are present in the sample (and, therefore, by the same proportion of defaulted observations to the total number of bad borrowers, and of performing observations to the total number of good borrowers).

In conclusion, the higher the AuROC, the better the discriminatory power of the variable. To obtain a ROC curve and AuROC for ROE, using SPSS-PASW (Figure 4.15):

Analyze, ROC curve, Test Variable(s):ROE, State Variable: 01Status, Value of state variable: 1, Display ROC curve e with diagonal reference line, Standard error e confidence interval, Options, Test direction, Smaller test indicates more positive results, Paste

The last option (smaller test) specifies that the variable must be used for ranking borrowers starting from its lowest values. In fact, it is necessary to start from ratio values indicating a higher default risk ('more positive results'); we already stated the working hypothesis for ROE by saying that lower ROE is associated with higher risk.

SPSS-PASW allows calculation of ROC curves and AuROC measures for many indicators all at once, provided that they all move in the same direction with default

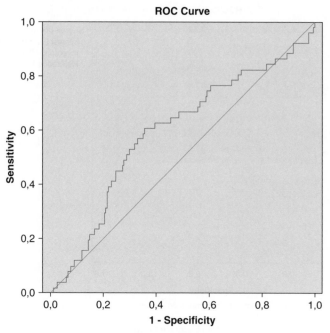

Diagonal segments are produced by ties.

Area Under the Curve

Test Result Variable(s):NetProfit/Equity

Area	Std. Error[a]	Asymptotic Sig.[b]	Asymptotic 95% Confidence Interval	
			Lower Bound	Upper Bound
0.593	0.043	0.025	0.509	0.676

The test result variable(s): NetProfit/Equity has at least one tie between the positive actual state group and the negative actual state group. Statistics may be biased.

a. Under the nonparametric assumption

b. Null hypothesis: true area = 0.5

Figure 4.15   ROC curve for ROE.

risk (i.e., the signs of their working hypotheses are the same). Figure 4.16 presents results for the other two variables that we are currently using as examples and that are moving in an opposite direction than ROE with respect to default risk.

These three ratios have AuROC measures that can be evaluated as average, considering that they are measuring discriminatory power of individual ratios and not of models as a whole. In addition, they have p-values far below 5% (the null hypothesis is that the true AuROC is 50%, therefore we expect to reject it by finding p-values lower than the set alpha-value).

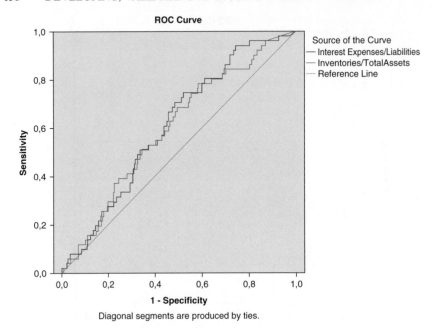

Diagonal segments are produced by ties.

Area Under the Curve

Test Result Variable(s)	Area	Std. Error[a]	Asymptotic Sig.[b]	Asymptotic 95% Confidence Interval	
				Lower Bound	Upper Bound
Interest Expenses/Liabilities	0.609	0.036	0.008	0.539	0.678
Inventories/TotalAssets	0.597	0.038	0.019	0.522	0.672

The test result variable(s): InterestExpenses/Liabilities, Inventories/TotalAssets has at least one tie between the positive actual state group and the negative actual state group. Statistics may be biased.

a. Under the nonparametric assumption

b. Null hypothesis: true area = 0.5

Figure 4.16   ROC curves for two ratios.

When looking at Figure 4.15, it is evident that ROE is not capable of correctly identifying defaults when its values are very low or very high. In these extreme regions, the ROC curve performs worse than the casual straight line indicating random selection. It is a consequence of the many problems of ROE which have been already observed.

Consider all variables in the long list. Almost all of them have AuROC measures higher than 50%; this is a confirmation that working hypotheses previously set are aligned with empirical evidence as shown by ROC curves. Unexpected results concern three variables. For IeOnLiablities and Debt_Equity_Ratio, in theory, a positive relationship with default was expected: empirical evidence read through ROC curves cannot change the theory, but can suggest that they are problematic indicators, probably because of their structures and underlying accounting practices. For Payables_Period a negative relationship with default risk was expected given that higher values of the ratio are usually seen as indicating less financial needs and stronger negotiating power with suppliers. However, we have already outlined that this ratio can also be interpreted as an expression of a firm's financial tensions that ask for longer payment terms to suppliers. In this case, empirical evidence can be supported by the latter theoretical interpretation of the ratio.

Far more than 50% of the ratios have a statistical significance of AuROC at 5%. Note that the ratios which have AuROC less than 59% do not present statistical significance of the measure; the three ratios with unexpected results are among them.

ROC Curves indicate that ratios which have the best results (all statistically significant at 5%) are, in descending order, based on AuROC: Debt/EquityTraditional ratio (67%), EBITDAonIE (65%), AssetsTurnover (64%), Inventory_Period (63%) and all the following Commercial_WC_Period, IEonEBITDA, ROAminusIEonTL sharing an AuROC of 62%. These ratios are natural candidates to enter into a statistical-based model which aims to predict default risk.

4.5.6   Empirical monotonicity

Structural monotonicity has a twin concept based on empirical observation. We have already stated that empirical monotonicity is satisfied when categorizing the variable into contiguous ordered intervals, and calculating the average default rates of borrowers included into such intervals we obtain a monotonic increase (or decrease) of default rates. Availability of empirically monotonic explanatory variables helps to build robust statistical based rating systems. If a variable is not empirically monotonic, it can be considered convenient to transform it. This kind of transformation is, however, not simple and is examined later.

To assess empirical monotonicity of a variable, for example ROE, we can use the SPSS-PASW main menu. Suppose we want to create a new variable (ROE_bin9) that categorizes ROE into ten equally populated bins:

> *Transform, Visual Binning, ROE, Continue, Binned variable: ROE_ bin9, Make cutpoints, Equal percentiles based on scanned cases, Number of cutpoints: 9, Apply, Make labels, Paste*

After running the syntax, the new variable will be added to the dataset. It is necessary to specify in the variables view that the measure type of the new variable is nominal. Now, we can obtain a table which reports default rates for each bin of ROE_bin9:

> *Analyze, Tables, Custom Tables,* move 01Status to the horizontal bar and ROE bin_9 to the vertical bar, then, with the mouse pointer on 01Status right click to activate *Summary Statistics,* choose *Mean, Apply to selection, Paste,* then *Run* the selected syntax.

The results indicate that ROE is not empirically monotonic when ten bins are considered. In order to draw a graphical representation of the relation between ROE bins and their default rate, do the following:

Graph, Chart builder, Gallery: Line, drag and drop *ROE_bin9* to box as X, do the same for 01Status as Y (it is necessary that the variable is considered as scale, if it is not, right click on its label in the graph and choose *Scale*), *Element properties, Line1, Statistics: mean*

Running the syntax produces Figure 4.17.

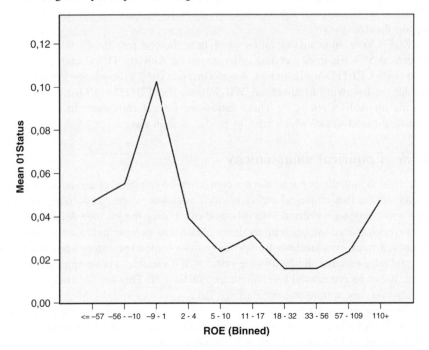

Figure 4.17   Default rates for ROE intervals.

There are many possible factors explaining this strange dynamic of default rates for contiguous ordered intervals of ROE. Some factors are linked to firms' peculiar situations that are misled by the ratio. For instance, a large ROE can be derived from a small equity, perhaps eroded by many years of losses; firms in this situation may have high ROEs but are deemed as being high risk. The absence of structural monotonicity may also derive from the possibility to have negative denominators (Total Stockholders' Equity is net of loans to shareholders).

To verify the impact of the last factor, let's observe numerator and denominator signs and their combinations. From the SPSS-PASW main menu:

Transform, Recode into different variables, Input variable: Equity, *Output variable:* EquitySign, *Change, Old and new values, Range lowest through value: 0, New value: − 1, Add, Range value through highest: 0, New value: 1, Add, Value: 0, New value: 0, Add, Continue, Paste*

Transform, Recode into different variables, Input variable: NPROF, *Output variable:* NPROFsign, *Change, Old and new values, Range lowest through value: 0, New value: − 1, Add, Range value through highest: 0, New value: 1, Add, Value: 0, New value: 0, Add, Continue, Paste*

After running the syntax and obtaining the categorized variables, save the new dataset as:

W_File, save as: W_CS_1_AnalysisSampleDataSet_3x

Then, select a cross tabulation by:

Analyze, Descriptive Statistics, Crosstabs, row EquitiSign, *column* NPROFsign, *Layer:* 01Status, *Cells, Counts: Observed, Percentage: Total, Continue, Paste, Run to end*

Table 4.18 reports the results. The denominator is negative for 200 cases, usually associated with a positive net profit and performing borrowers. In 51 cases, ROE is positive as a result of negative signs for both numerator and denominator. In conclusion, empirical non-monotonicity can be determined by both economic reasons and weaknesses of the ratio's structure.

Table 4.18   Distribution by numerator and denominators signs.

EQUITYSigns * NPROFSigns * 01STATUS Cross-tabulation						
01STATUS				NPROFSigns		Total
				−1.00	0.00	1.00	
performing	EQUITYSigns	−1.00	Count	47	2	138	187
			% of Total	3.8	0.2	11.3	15.3
		0.00	Count	0	0	6	6
			% of Total	0	0	0.5	0.5
		1.00	Count	165	14	849	1028
			% of Total	13.5	1.1	69.5	84.2
	Total		Count	212	16	993	1221
			% of Total	17.4	1.3	81.3	100.0
default	EQUITYSigns	−1.00	Count	4		9	13
			% of Total	7.8		17.6	25.5
		1.00	Count	14		24	38
			% of Total	27.5		47.1	74.5
	Total		Count	18		33	51
			% of Total	35.3		64.7	100.0

Empirical monotonicity of ROE has been verified using 10 intervals and calculating default rates for each interval. Lowering the number of bins increases the probability of reaching empirical monotonicity, and vice versa. Therefore, this characteristic is relative; if the number of bins is increased enough, no ratio will be empirically monotonic. In our sample, only some ratios are empirically monotonic, if categorized by a reasonable number of cut-offs, and are as following: EBITDAonSales, ROI, Assets_Turnover, Inventory_Period, Payables_Period, IEonLiablities, IEonFinancial_Debts, Trade_PayablesOnTL, InventoriesOnTA, EquityonPermanent_Capital, ROAminusIEonTL. Other ratios should be transformed in order to meet the requirement.

4.5.7   Correlations

It is useful to examine the correlation between pairs of variables because:

(a) high correlations can lead to stability problems in the estimation of coefficients of the scoring functions;

(b) highly correlated indicators illustrate similar information and their presence complicate and confuse the analysis rather than being helpful.

For these reasons, it is advantageous to use low correlated indicators wherever possible. Erasing the variable with the lower discriminatory power in each highly correlated pair leads to an efficient short list of explanatory variables for the development of multivariate models. The literature does not indicate any compulsory limit. In fact, the relevant triggering level of alert depends on the specific application of statistical tools and the nature of data. For data originating from balance sheets, that is to say from accounts that must balance, it is normal to have levels of correlation that would be considered very high for other applications, such as those concerning sociological phenomena, for instance. In addition, different statistical indicators of correlation lead to different levels of correlation coefficients by themselves; we consider Pearson's parametric correlation and Spearman's non-parametric (rank ordering) correlation, by:

Analyze, Correlate, Bivariate, Variables: all ratios, then select *Pearson* and *Spearman, Flag significant correlation, Paste*

Running the syntax produces an output reporting correlations matrix, correlation coefficients, p-values of Student's t-test and the number of observations. The hope is to have p-values less than 0.05 to conclude that the calculated correlation coefficient has not been obtained by chance. Save the files as:

W_AnalysisSampleOutput_5.spv

W_AnalysisSampleSyntax_5.sps

Consider the usual three variables in Table 4.19. When pairing only these three variables, there is a unique case of statistically significant correlation (at 1%).

Table 4.19   Correlation.

		NetProfit/ Equity	InterestExpenses/ Liabilities	Inventories/ TotalAssets
NetProfit/Equity	Pearson Correlation	1	−0.025	−0.034
	Sig. (2-tailed)		0.366	0.220
	N	1272	1272	1272
InterestExpenses/ Liabilities	Pearson Correlation	−0.025	1	−0.030
	Sig. (2-tailed)	0.366		0.282
	N	1272	1272	1272
Inventories/ TotalAssets	Pearson Correlation	−0.034	−0.030	1
	Sig. (2-tailed)	0.220	0.282	
	N	1272	1272	1272

			NetProfit/ Equity	Interest Expenses/ Liabilities	Inventories/ TotalAssets
Spearman's rho	NetProfit/ Equity	Correlation Coefficient	1.000	−0.153[**]	−0.051
		Sig. (2-tailed)		0.000	0.069
		N	1272	1272	1272
	Interest Expenses/ Liabilities	Correlation Coefficient	−0.153[**]	1.000	0.021
		Sig. (2-tailed)	0.000		0.448
		N	1272	1272	1272
	Inventories/ TotalAssets	Correlation Coefficient	−0.051	0.021	1.000
		Sig. (2-tailed)	0.069	0.448	
		N	1272	1272	1272

[**]Correlation is significant at the 0.01 level (2-tailed).

It concerns Spearman's correlation between IEonLiabilities and ROE; it has a negative sign (as expected) and a correlation coefficient level quite low, indicating that the strength of correlation is also low.

Correlation results for the whole set of variables in the long list are summarized in the univariate analyses summary table we will examine later. Moving from parametric correlation (Pearson's) to non-parametric correlation (Spearman's), more pairs of variables denote higher coefficient levels as well as statistically significant correlations (at 5%). On the other hand, the number of correlations which have a coefficient beyond 70% is similar, but variables involved are often different. These differences between Pearson's and Spearman's correlations depend on the different nature of correlation they consider. The former is based on scale values of financial indicators; the latter is based on ranked values of variables and it is particularly useful when non-normal indicators, many outliers and small samples are involved.

Very high correlation coefficients are observable (among others) between the following variables:

- IEonEBITDA and NIEonEBITDA, 99%

- EBITDA/Sales and EBIT/Sales (ROS), 99.4%

- ROA and ROI, 99.7%

- ROA and AssetsTurnover, 98.9%

- ROI and ROAminusIEonTL, 99.7%

- AssetsTurnover and ROAminusIEonTL, 98.9%

- ROS and EBITDAonVP, 92.8%.

It should be also pointed out that the negative correlation between DebtEquityTr and ROEtr is just below 70%.

In addition, it should be noted that empirical evidence concerning unexpected relative positions of means in subsamples of good and bad borrowers, non-significant association tests, non-normality, lack of homogeneity of variance, poor discriminatory power, and non-monotonicity of variables, may all depend on the small sample size and in particular, on the small number of defaults. Other reasons concern the chosen structure of financial ratios, the small size of borrowing firms and their sometimes quirky accounting. Some of these problems can be addressed by analyzing and managing outliers and by appropriate ratios transformations.

4.5.8   Analysis of outliers

When there are values that are considerably different from the others, they can strongly impact on the suitability of the indicator to enter into a statistical-based model. The outliers bias the mean, inflate the standard deviation, and affect all subsequent analyses. We need to know how many observations present a relevant number of outliers, which observations are concerned, and how many ratios are strongly impacted by outliers. To perform these analyses there are three alternative approaches.

The first approach: if there is a clear idea about values of a given variable to be considered 'normal', analyzing percentiles can give a clue on how many observations have abnormal values. In this case, canonical definitions of outliers are not used to identify, quantify and manage them. Instead, a case-by-case strategy is developed.

The second approach: standardize financial ratios by subtracting the ratio's mean and dividing by its standard deviation. Standardized values (named Z-score by SPSS-PASW) have two characteristics: their mean is zero, and their unit of measure is the standard deviation. A classic definition of outliers is that of values whose Z-scores are external to the ± 3.29 interval. In fact, if a variable is normally distributed, it is expected that relationships reported in Table 4.20 are respected.

Table 4.20   Percentiles for different (absolute) Z-scores.

Percentiles of observations		Z-scores
95	have (absolute) Z-scores lower than	1.96
5	have (absolute) Z-scores higher than	1.96
1	have (absolute) Z-scores higher than	2.58
0	have (absolute) Z-scores higher than	3.29

To calculate ratios' standardized values by SPSS-PASW, select the following from the main menu:

> *Analyze, Descriptive Statistics, Descriptives,* flag *Save standardised values as variables,* all ratios, *Options,* choose *Mean* (at least one statistic is required).

A weakness of this approach is that observations' values are defined as outliers on the basis of outlier values themselves, because extreme values impact on standard deviation, which is used to identify outliers.

The third approach: outlier definition can be based upon multiples of interquartile range plus/minus respectively the upper or lower bound of interquartile range. For instance, outliers are values higher than 3rd quartile $+3 \times$ IQR, or values lower than 1st quartile $-3 \times$ IQR. In this analysis we do not differentiate between good and bad borrowers. To identify outliers using this approach it is convenient to export the dataset into Excel. From the SPSS-PASW main menu:

> *File, Save as, Save as Type:* desired Excel release, *File name:* Out, *Save*

Check that Excel is using the same international format for numbers as SPSS-PASW. Insert a second worksheet and name it LowerBound. Copy all financial ratios names (starting from ROE) present in the first row of the Excel worksheet created by SPSS-PASW and paste them into the second worksheet denominated LowerBound. Then in cell A2, type:

> *=percentile(Out!ch2:ch1273;0,25)−3*(percentile(Out!ch2:ch1273;0,75)*
> *−percentile(Out!ch2:ch1273;0,25))*

CH is the column of ROE in the Excel worksheet created by SPSS-PASW, named as the file. Now, copy this formula and paste it for all other financial ratios on the right.

Insert a third worksheet and name it UpperBound, duplicate the same copy/paste procedure as outlined above, and type in cell A2:

=percentile(Out!ch2:ch1273;0,75)+3*(percentile(Out!ch2:ch1273;0,75)
−percentile(Out!ch2:ch1273;0,25))

Subsequently, insert a fourth worksheet and name it CountOut. Paste BOR-CODE and 01Status data in the first two columns, using the first row for variables names. From cell C1 to the right end of the first row paste financial ratios names, starting from ROE and ending with ROEtr; in cell C3, type the following formula:

=If(O(Out!CH2<LowerBound!A$2;Out!CH2>UpperBound!A$2);1;0)

Then, copy the formula to the right side for all variables and down for all observations. In the second row, calculate the sum of all cells below; it will be the number of outliers per ratio. In the first empty column to the right calculate the sum of the row; it will show the number of outliers per observation.

There are only two ratios which have more than 10% of outliers (12% for ROE and 13% for Sales/VP).

The maximum number of outliers per borrower is 13 (taking into account all 34 ratios and also considering those already excluded from further analyses, such as Extr_I_C_onCurrent_Income and TaxesOnGross_Profit). Table 4.21 summarizes the results.

Table 4.21   Outliers per observation.

# outlier ratios	# observations
13	2
12	2
10	3
9	6
8	12
7	18
...	...
0	349

4.5.9   Transformation of indicators

4.5.9.1   Treatment of outliers and shapes of distributions

Outliers' analyses are conducted both for each observation and for each ratio. The former requires setting a limit for the number of ratios with more than a given number of outliers, beyond which the observation should be erased from the dataset (for instance, one third of ratios), possibly after having checked that there are no mistakes in the data collection step. In our case study we will save all observations.

The latter indicates that the number of outliers per ratio is almost always less than 10%. Therefore, substituting them with lower bounds and upper bounds previously determined could be an acceptable solution. This substitution process is known as 'winsorization'.

As we have argued before about substituting missing values with means or medians, we can choose to calculate substitution values separately for the two subsamples of good and bad borrowers in the analysis sample by using two different limits for the two groups. However, for the sake of simplicity, we are going to use single lower and upper limits for each ratio in developing our case study. It is useful to use the same Excel file where the number of outliers was counted. Insert a fifth worksheet (to be named '3iqr') and paste the names of ratios from ROE to ROEtr on the first row; type in cell A2 the following formula:

$$=If(Out!CH2<LowerBound!A\$2;LowerBound!A\$2;If(Out!CH2>$$
$$UpperBound!A\$2;UpperBound!A\$2; Out!CH2))$$

Now, copy this cell and paste it down and on the right until all cells (for all observations and all variables) are covered. Save the file and then open a new empty .sav file in SPSS-PASW. From the variable view of *W_CS_1_ AnalysisSampleDataSet_3*.sav file copy columns from 'name' to 'measure' for all ratios from ROE to ROEtr. Now select the first 34 rows of the variable view in the new empty file and paste it. Edit variable names by highlighting them and pressing F2, typing 3 to indicate that these variables have been constrained within three times the interquartile range from the boundaries of the IQR. Go back to the Excel file, worksheet 3iqr: select variables values from A2 to AH1273 and copy them. Now, when you are again in the first cell of SPSS-PASW data view, *Edit, Paste*. Finally, save the new .sav file as:

W_ CS_1_AnalysisSampleDataSet_4

At this point, we need to merge W_CS_1_AnalysisSampleDataSet_4 and W_CS_1_AnalysisSampleDataSet_3. The easiest way is to copy the new 34 variables in the variable view of the former file (select from 'name' of the first variable to 'measure' of the last variable), and paste into the variable view of the latter file. Then, copy all available data from the data view of the former file (W_CS_1_AnalysisSampleDataSet_4), and paste it into the data view of W_CS_1_AnalysisSampleDataSet_3 (the area to be copied and pasted must be exactly the same size). The last step is to delete the labels of new variables, which are identical to that of the original variables, as it would create confusion in reading outputs. Save the file as:

W_CS_1_AnalysisSampleDataSet_5.sav.

Having added new variables to the dataset, we need to perform all descriptive analyses by the Explore routine again. We can re-load the syntax saved as *W_AnalysisSampleSyntax_2*. Add 3 to all variables names; then, save the new syntax as well as the two output files we obtain when running separately the two blocks of syntax:

W_AnalysisSampleSyntax_6.sps

W_AnalysisSampleOutput_6.spv

W_AnalysisSampleOutput_6plot e test.spv

We can also replicate ANOVA and ROC curves by modifying, as done above, the syntax in W_AnalysisSampleSyntax_3. Save the new files as:

W_AnalysisSampleOutput_6anova e roc.spv

W_AnalysisSampleSyntax_6anova e roc.sps

The simplified choice of using the same upper and lower limits for the distributions of good and bad borrowers has its disadvantages. The accumulation of observations on the limits set is evident. Nevertheless, distributions ranges appear to be much more reasonable. For instance, the ROE means of the two subsamples are now aligned with expectations. Vice versa, AuROC values are practically the same for the new and the corresponding old variables; in fact, ROC curves take account of ranks rather than of scale differences.

A different approach to manage outliers, which avoids accumulating observations on particular values of the distribution, is based on the logistic transformation of original financial indicators. It uses a continuous function, such as logistic function, which is naturally constrained between zero and one. Applying logistic (cumulative distribution) function to a financial ratio 'x' means transforming it according to the following formula:

$$y = \frac{1}{1 + e^{-(x-a)/b}}$$

where 'a' indicates the intercept (the height of 'y' when 'x' is zero) and 'b' indicates the slope of the function. Figure 4.18 depicts logistic functions which have different values of intercept and scale parameters.

If we want to increase the slope of the function and flatten its tails, we need to decrease the 'b' scale parameter. A higher slope increases the sensitivity of y to x in the central values of the distribution but, at the same time, cumulates more values of the transformed ratio near the asymptotic extremes of zero and one. Therefore, it is necessary to carefully set the scale parameter in order to balance the trade off.

An example of trials targeting a ROI transformation capable of transforming it in a normally distributed variable is shown in Figure 4.19. It is necessary to increase the value of the scale parameter 'b' in order to reduce the accumulation of observations in the tails of the transformed variable and to get closer to a normal distribution, as histograms and Q–Q plots clearly suggest. The optimal value of

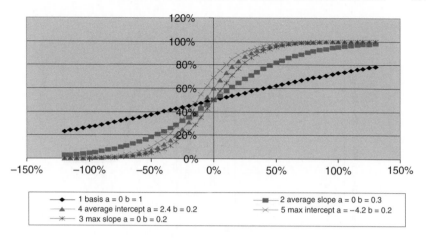

Figure 4.18   Logistic functions.

the scale parameter can change from financial ratio to financial ratio and therefore requires some refinement.

How can we use SPSS-PASW in order to transform explanatory variables and obtain graphs such as those in Figure 4.19? Focus on ROE; it has about 12% of outlier values, according to the definition of three times the IQR from IQR boundaries. A logistic transformation can help; start with an intercept value of zero and a scale value of one. Close all .spv and .sps files in case some of them are open, and use the dataset in *W_CS_1_AnalysisSampleDataSet_5.sav.* From the SPSS-PASW main menu:

> *Transform, Compute variable, Target variable:* T1ROE, *Function group: CDF and non central CDF, Functions and special variables: CDF logistic,* move up in the box and fill in the three question marks (representing *quant, mean, scale*) with the name of the variable to be transformed, intercept value and scale value: ROE, *0, 1, Paste, Run selection*

A new variable is added to the dataset. Save it as *W_CS_1_AnalysisSample DataSet_6.sav*

Produce descriptives for the original variable ROE and for the transformed variables ROE3 and T1ROE, without factoring performing and defaulted borrowers:

> *Analyze, Descriptive Statistics, Explore, Dependent list: ROE ROE3 T1ROE, Both, Statistics: descriptives, percentiles, Continue, Plots: Factor level together, Normality plot with test, Continue, Paste, Save as: W_AnalysisSampleSyntax_7trasf.sps, Run selection*

Logistic transformation has worsened the situation, accumulating observations in the tails of the distribution of the variable transformed by the logistic function.

168   DEVELOPING, VALIDATING AND USING INTERNAL RATINGS

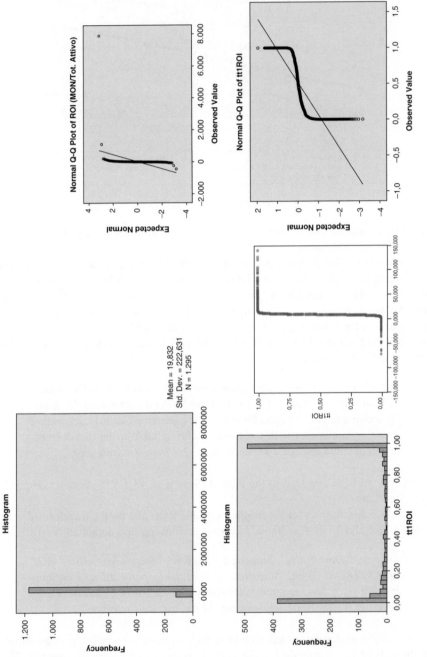

Figure 4.19   Transforming ROI.

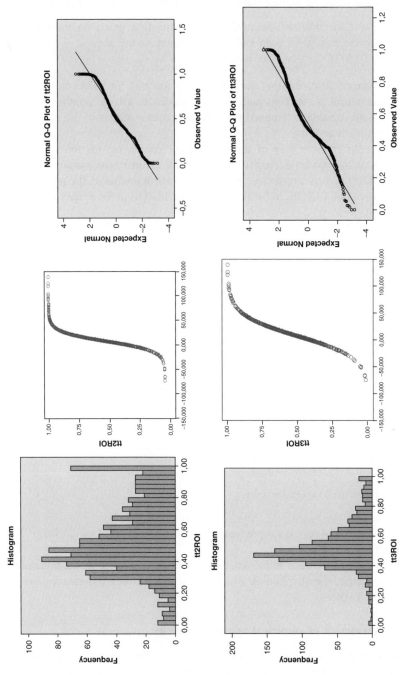

Figure 4.19   (continued)

Perhaps the slope parameter is too low and the function slope is too high. To get an idea of the shape of logistic transformation, plot ROE against T1ROE by:

> *Graph, Chart Builder, Scatter/dot,* move the first graph on the top-left to the box, then drag and drop variables to the axes $X = ROE$ $Y = T1ROE$, *Paste, Run selection*

We obtain confirmation that the slope of the function for values in the middle of the distribution is too high and, as a consequence, a large number of values of the original variable is flattened towards the extreme values of the transformed distribution (zero and one).

To have an efficient way of testing different parameters for the slope, copy the syntax of logistic transformation and paste it again in the syntax file. Now, substitute 1500 to 1 as slope value in the pasted script and name the new variable T2ROE instead of T1ROE. Substitute T2ROE to T1ROE in the syntax of all other analyses:

```
DATASET ACTIVATE DataSet1.
COMPUTE T2ROE = CDF.LOGISTIC(ROE,0,1500).
EXECUTE.
EXAMINE VARIABLES = ROE ROE3 T1ROE T2ROE
/PLOT BOXPLOT HISTOGRAM NPPLOT
/COMPARE GROUPS
/PERCENTILES(5,10,25,50,75,90,95) HAVERAGE
/STATISTICS NONE
/MISSING LISTWISE
/NOTOTAL.
DATASET ACTIVATE DataSet1.
* Chart Builder.
GGRAPH
/GRAPHDATASET NAME = 'graphdataset' VARIABLES = ROE T2ROE MISSING =
LISTWISE REPORTMISSING = NO
/GRAPHSPEC SOURCE = INLINE.
BEGIN GPL
SOURCE: s = userSource(id('graphdataset'))
DATA: ROE = col(source(s), name('ROE'))
DATA: T2ROE = col(source(s), name('T2ROE'))
GUIDE: axis(dim(1), label('ROE'))
GUIDE: axis(dim(2), label('T2ROE'))
ELEMENT: point(position(ROE*T2ROE))
END GPL.
```

Save files *W_AnalysisSampleSyntax_7trasf.sps* and *W_AnalysisSampleSyntax_7trasf.spv*.

The situation has much improved. Much like we saw for ROI, the transformation has re-shaped the distribution, which is now much closer to normality and less impacted by outliers.

4.5.9.2 Treatment of empirical non-monotonicity

A solution often used to overcome the issue of empirical non-monotonic relationships with probabilities of default is the transformation of indicators into probabilities of default. Contiguous intervals in the indicators' values are chosen, and the default rate for each interval is empirically determined on the basis of the analysis sample. The average indicator value for an interval is named a 'node', and the calculated default rate of the interval is associated with it. Then, these default rates are connected by interpolation.

In Figure 4.20, 'K' and 'TK' represent the values of the un-transformed and transformed financial indicators. For each couple of intervals, 'u' and 'l' represent the upper and lower limits of the transformation calculated as empirical default rates of the corresponding intervals. They are associated to the average value of K for each interval (nodes).

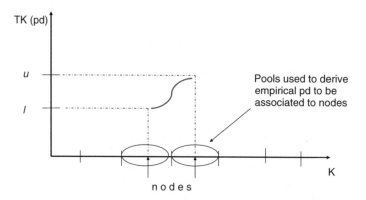

Figure 4.20   Transforming a financial indicator into a default rate.

The result of the transformation described above is the assignment of a sample-based default probability to each possible indicator value. This form of variable transformation in sample-based probabilities of default is a solution for both the problem of empirical non-monotonicity and the problem of excessive outliers. However, it can introduce some subjectivity when the choice of intervals to calculate nodes and default rates is discretionary. In addition, it is, in any case, important to investigate hypothesis violations using the untransformed indicator, because every indicator will meet the conditions of the working hypothesis after transformation into a default probability.

Nonlinear interpolation is preferable to smooth out the change of direction of the transformation function. A possible formula to use is as follows:

$$TK = l + \frac{u-l}{1+e^{a-bK}} = l + (u-l) \times \frac{1}{1 + \frac{1}{e^{-a+bK}}}$$

A logistic function is used to connect nodes. The parameters 'b' and 'a' respectively represent the slope of the curve and the location of the inflection point. K is not exactly the original variable, but a transformation of it that allows the logistic function to get closer to the zero-one extremes.

To calculate K intervals and the default rates for each interval (to be associated with the node's values), the same procedure used to check ROE empirical monotonicity can be used: we need to categorize ROE based on a chosen number of classes and, then, to calculate the default rates for each class. Excel can be one of simplest tools for setting 'a' and 'b' parameters and applying logistic functions to link default rates of each contiguous pairs of nodes.

4.5.9.3   Other transformation: categorical variables

Categorical variables may also require transformations when, for instance, the number of observations for different levels of the variable is too low, or when the number of missing values in some levels is too high. In fact, a possible solution to avoid excluding these variables is to aggregate levels (for instance, subsectors or even sectors). Consider sectors. Are they well populated in the sample? To get the answer, select the following from SPSS-PASW main menu:

> *Analyze, Descriptive statistics, Crosstabs, Rows: Sector, Columns: 01Status, Cells, Counts: Observed, Percentages: Row e Total, Continue, Paste, Run selection*

Results show that many sectors are poorly populated. We can define a policy for our analysis stating that the minimum number of defaults per sector should be five and then we can observe how many sectors have a default rate different enough from the general mean (4%). There are only three sectors meeting these requirements: #430 with a default rate of 4.4%, #614 with 6.6% that are above the average, and #482 that shows a default rate of 2.9% well below the average. All other sectors could be aggregated in a unique residual category by:

> *Transform, Recode into different variables, Numeric variable: Sector, Output variable: Sector3Cat, Change, Old and new variables: 430->430614, 614->430614, 482->482, All other values->0, Continue, Paste,*

By selecting the same cross-tabulation as before for the new variable, we obtain Table 4.22. Results suggest the opportunity to use only two categories, because the residual category created has a default rate very close to that of sector #482.

As this new variable has only two levels, it can be considered as a dummy variable, facilitating its direct use in multivariate procedures. From SPSS-PASW main menu, select:

> *Transform, Recode into different variables, Numeric variable: Sector, Output variable: OverSect430_614, Change, Old and new variables: 430->1, 614->1, All other values->0, Continue, Paste, Run selection.*

A last aspect to debate for categorical variables is the opportunity to generate them from scale variables. For example, you may want to analyze whether a firms size impacts the probability of default. Start by using an original scale variable, such as Sales. We can compare firms' sales with the nominal 01Status variable

Table 4.22   Default rates for recoded sectors.

SECTOR3cat * 01STATUS Cross-tabulation

			01STATUS		
			performing	default	Total
SECTOR3cat	0	Count	202	6	208
		% within SECTOR3cat	97.1	2.9	100.0
		% of Total	15.9	0.5	16.4
	482	Count	236	7	243
		% within SECTOR3cat	97.1	2.9	100.0
		% of Total	18.6	0.6	19.1
	430614	Count	783	38	821
		% within SECTOR3cat	95.4	4.6	100.0
		% of Total	61.6	3.0	64.5
Total		Count	1221	51	1272
		% within SECTOR3cat	96.0	4.0	100.0
		% of Total	96.0	4.0	100.0

by many of the methods already examined to verify the discriminatory power of explanatory variables.

By adopting ROC curves, we can obtain results that are aligned to those of many ratios (Figure 4.21). But, if we give a glance to the Independent t-test and its related measure of association (Table 4.23), we realize that the relationship is not statistically significant and the strength of the relationship is very low (0.04).

Evaluate if it is beneficial to categorize the sales scale variable in classes by:

Transform, Visual Binning, Variable to bin: Sales, *Continue, Binned variable:* SalesBin4, *Make cutpoints, Equal percentiles: 4, Apply, Make labels, Paste*

Copy and paste the cross-tabulation syntax we got before and substitute *Sales-Bin4* to *Sector3Cat* in the script; then, Run selection. The low number of defaults in the first class and the proximity of its default rate to those of the subsequent three classes suggest that we should verify if it would be better to have a first class reaching 1000 and a last class starting from 7500 (remember that data are in thousands of Euro). Let's re-apply the visual binning routine selecting only two cut-points, naming the new variable as SalesBin2 and then modifying class' boundaries as mentioned above (before clicking on make labels).The following syntax will be created:

```
* Visual Binning.
*Sales.
RECODE Sales (MISSING = COPY) (LO THRU 1000.0 = 1) (LO THRU 7500.0 =
2) (LO THRU HI = 3) (ELSE = SYSMIS)
INTO SalesBin2.
VARIABLE LABELS SalesBin2 'Sales (Binned)'.
```

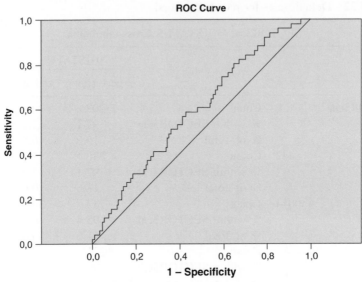

Diagonal segments are produced by ties.

Area Under the Curve

Test Result Variable(s):Net sales

Area	Std. Error[a]	Asymptotic Sig.[b]	Asymptotic 95% Confidence Interval	
			Lower Bound	Upper Bound
0.598	0.039	0.018	0.522	0.673

The test result variable(s): Net sales have at least one tie between the positive actual state group and the negative actual state group. Statistics may be biased.
[a] Under the nonparametric assumption
[b] Null hypothesis: true area = 0.5

Figure 4.21   ROC curve of Sales.

```
FORMATS SalesBin2 (F5.0).
VALUE LABELS SalesBin2 1 '< = 1000' 2 '1001 - 7500' 3 '7501+'.
MISSING VALUES SalesBin2 ().
VARIABLE LEVEL SalesBin2 (ORDINAL).
EXECUTE.
```

Produce the usual cross-tabulation on the new variable, now also including the test of association (Chi-squared) and the measure of association (Cramer's V, as the contingency table would have more than 2×2 dimensions). Table 4.24 and Table 4.25 indicate that now we have three well populated classes and default rates which appear to be well differentiated. The statistical test confirms that differences are statistically significant at 5%. Even if the two variables have been considered as nominal, in any case the contingency table should be read in terms

Table 4.23 Independent t-test for Sales.

Group statistics

	01STATUS	N	Mean	Std. Deviation	Std. Error Mean
Net sales	performing	1221	7461.42	22 641.492	647.959
	default	51	12 218.82	29 309.882	4104.204

Independent samples test

		Levene's test for equality of variances		t-test for equality of means						95% Confidence Interval of the Difference	
		F	Sig.	t	df	Sig. (2-tailed)	Mean Difference	Std. Error Difference		Lower	upper
Net Sales	Equal variances assumed	3.341	0.068	−1.451	1270	0.147	−4,757.399	3,278.744		−11,189.749	1,674.951
	Equal variances not assumed			−1.145	53	0.257	−4,757.399	4,155.038		−13,093.121	3,578.323

Table 4.24   Cross-tabulation for binned sales.

SALES (Binned) * 01STATUS Cross-tabulation

			01STATUS performing	default	Total
SALES (Binned)	<= 1000	Count	380	9	389
		% within SALES (Binned)	97.7	2.3	100.0
		% of Total	29.9	0.7	30.6
	1001–7500	Count	601	26	627
		% within SALES (Binned)	95.9	4.1	100.0
		% of Total	47.2	2.0	49.3
	7501+	Count	240	16	256
		% within SALES (Binned)	93.8	6.3	100.0
		% of Total	18.9	1.3	20.1
Total		Count	1221	51	1272
		% within SALES (Binned)	96.0	4.0	100.0
		% of Total	96.0	4.0	100.0

Table 4.25   Chi-square tests and measures of association for binned sales.

Chi-square tests

	Value	df	Asymp. Sig. (2-sided)
Pearson Chi-Square	6.277	2	0.043
Likelihood Ratio	6.312	2	0.043
Linear-by-Linear Association	6.257	1	0.012
N of Valid Cases	1272		

Symmetric measures

		Value	Approx. Sig.
Nominal by Nominal	Phi	0.070	0.043
	Cramer's V	0.070	0.043
N of Valid Cases		1272	

[a]0 cells (0.0%) have expected count less than 5. The minimum expected count is 10.26.

of coherence with mainstream expected relation between firms' size and probability of default. It is commonly observed that default rates decrease when firms'
size increases. However, the sales classes we have created and the sample as a
whole both refer to very small firms. Therefore, the obtained result does not *per
se* contradict common beliefs.

Up to now, we have three levels of the new categorical variable; to consider it in the short list for the multivariate procedure, it is necessary to build $3 - 1 = 2$ dummy variables. When both of these dummy variables assume a value of zero, the multivariate function will adopt the level of the variable that has been excluded as the baseline.

Using the transform routine, which has already been used many times, we can create the dummy variables concerning the first two levels of SalesBin2:

```
RECODE SalesBin2 (1 = 1) (ELSE = 0) INTO SalesBin2cl1.
EXECUTE.
RECODE SalesBin2 (2 = 1) (ELSE = 0) INTO SalesBin2cl2.
EXECUTE.
```

Running the syntax will create the new two variables in .sav file. Save files as:

W_AnalysisSampleSyntax_8trasfCat.sps

W_AnalysisSampleOutput_8trasfCat.spv

W_CS_1_AnalysisSampleDataSet_7.

4.5.10   Summary table of indicators and short listing

Univariate analysis on individual indicators may conveniently be summarized in a synoptic table. It gives an overview of variables' properties and allows us to make decisions such as:

(a) accepting the indicator in the short list of variables to submit to multivariate analyses,

(b) transforming the indicator to solve its problems in order use it in multivariate analyses,

(c) not including the indicator in the short list.

When some indicators are transformed, for instance to solve problems concerning outliers, it is necessary to replicate univariate and bivariate analyses for the new variable and fill in a new row of the summary table.

Table 4.26 reports an example of a table summarizing results for the 'analysis dataset' used in this case study. In the book, analyses reported have often been developed and commented on only for three ratios. One can use the summary table to check analyses performed on other variables and can also add further variables, obtained by other transformations of original indicators.

In fact, the many examined options can lead to many diverse choices (such as those concerning outliers' identification and management, or rather those concerning transformation of variables). Also, the final decision on whether to enter a variable into the short list is in part subjective, as different indicators may present opposite strengths and weaknesses and because different techniques may lead to different results for the same variable. Note, for instance, that AuROC is statistically

Table 4.26 Summary table of univariate and bivariate analyses.

# Ratio	It has economic meaning (Y/N); sign of relation with PD (+,-)	Structural and Empirical Monotonicity	Means. Trimmed means and Medians are aligned with hypotheses	Histograms and Q–Q plot suggest normality of distributions Good Bad subsamples	Normality test (K-S; SW) is not significant (p-value > 0.05) Good Bad subsamples	Homogeneity of Variance (Levine's) test is not significant (p-value > 0.05)	Means difference t-test is significant (p-value < 0.05)	Means difference F-test is significant (p-value < 0.05)	F-value	AuROC value Statistical significance at 5%	Correlation > 70% Pearson's *Spearman's*	# Outliers (3 × IQR)	Conclusions
Legend	Y/N + −	Y/N Y/N	Y/N Y/N	Y/N Y/N	Y/N Y/N	Y/N	Y/N	Y/N	Value	Value Y/N	Ratios	Number	
1 ROE	Y (−)	N N	N N Y	N N	N N / N N	N	N	Y	14.9	59 Y	ROEtr	154	
2 EBITDAon SALES	Y (−)	Y Y	N Y Y	N N	N N / N N	Y	N	N	0.0	57 N	EBITDA onVP *EBITDA onSALES* *EBITDA onVP* *EBITDA onSALES*	40	
3 ROI	Y (−)	Y Y	N Y Y	N N	N N / N N	N	N	Y	21.9	59 Y	Assets_Turnover IEon LIABILITIES ROAminus IEonTL *ROA* *MOLsuVPr* *ROS* *ROAminus* *IEonTL*	45	

4 ROA	Y (−)	Y	N	N	Y	Y	N	N	N	N	N	N	Y	22.2	58	Y	Assets_Turnover, IEon, LIABILITIES, *ROAminus*, *IEonTL*, *ROEtr*, *ROI*	63
5 EBITDA onVP	Y (−)	Y[1]	N	N	Y	Y	N	N	N	N	Y	N	N	0.0	59	Y	ROS, EBITDAonSALES, *ROS*, *ROI*, *EBITDA*, *onSALES*	35
6 ROS	Y (−)	Y	N	N	Y	Y	N	N	N	N	Y	N	N	0.0	57	N	EBITDAonSALES, ROA, EBITDAonSALES, IEon, LIABILITIES, *EBITDA*, *onSALES*, *ROI*, *EBITDA*, *onVP*	51
7 ASSETS_TURNOVER	Y (−)	Y	Y	N	Y	Y	N	N	N	N	N	Y	Y	23.0	64	Y	ROAminus, IEonTL, IEon, LIABILITIES, ROI	16
8 INVENTORY_TURNOVER	Y (−)	Y	N	Y	Y	Y	N	N	N	N	Y	N	N	0.3	54	N	ROI	126

(continued)

Table 4.26   (continued)

# Ratio	It has economic meaning (Y/N); sign of relation with PD (+,−)	Structural and Empirical Monotonicity	Means. Trimmed means and Medians are aligned with hypotheses	Histograms and Q–Q plot suggest normality of distributions Good Bad subsamples	Normality test (K-S; SW) is not significant (p-value > 0.05) Good Bad subsamples	Homogeneity of Variance (Levine's) test is not significant (p-value > 0.05)	Means difference t-test is significant (p-value < 0.05)	Means difference F-test is significant (p-value < 0.05)	F-value	AuROC value Statistical significance at 5%	Correlation > 70% Pearson's Spearman's	# Outliers (3 × IQR)	Conclusions
Legend	Y/N + −	Y/N Y/N	Y/N Y/N Y/N	Y/N Y/N	Y/N Y/N	Y/N	Y/N	Y/N	Value	Value Y/N	Ratios	Number	
9 RECEIVABLES TURNOVER	Y (−)	Y N	Y Y Y	N N	N N	Y	N	N	0.3	55 N		117	
10 RECEIVABLES PERIOD	Y (+)	Y N	N Y Y	N Y	N N Y Y	Y	N	N	0.0	57 N	INVENTORY_PERIOD COMMERCIAL_WC_PERIOD TRADE_RECEIVABLES onTA	19	
11 INVENTORY PERIOD	Y (+)	Y Y	N Y Y	N N	N N N N	Y	N	N	0	63 Y	COMMERCIAL_WC_PERIOD RECEIVABLES_PERIOD INVENTORIES onTA	71	

No.	Variable	Sign											Stat				Correlated variables
12	PAYABLES_PERIOD	Y (−)	Y	Y	Y	Y	N	N	Y	Y	N	N	0.4	48	N	47	
13	COMMERCIAL_WC_PERIOD	Y (+)	Y	N	N	Y	Y	N	Y	Y	N	N	0.0	62	Y	48	RECEIVABLES_PERIOD, INVENTORY PERIOD
14	IEon EBITDA	Y (+)	N²	Y	Y	Y	Y	N	N	Y	N	N	0.5	62	Y	65	NIEon EBITDA, NIEon EBITDA
15	NIEon EBITDA	Y (+)	N	N	Y	Y	Y	N	N	Y	N	N	0.5	60	Y	62	IEonEBITDA, IEonEBITDA
16	IEon LIABLITIES	Y (+)	Y	Y	Y	Y	Y	N	N	N	N	Y	23.7	61	Y	5	ROAminus, IEonTL, ROI, ASSET_TURNOVER
17	IEonFINANCIAL_DEBTS	Y (+)	Y	N	N	N	N	N	N	Y	N	N	0.5	48	N	60	
18e	EXTR_LC_onCURRENT_INCOME	Y (?)[3]	N	N	?	?	?	N	N	Y	N	N	0.4	?		310	
19e	TAXESon GROSS_PROFIT	Y (?)[3]	N	N	?	?	?	N	N	N	N	N	0.1	?		55	
20	INTANGIBLESonTA	Y (+)	Y	N	Y	Y	Y	N	N	Y	N	N	0.0	52	N	0	
21	TRADE_RECEIVABLES onTA	Y (−)	Y	N	Y	Y	Y	N	Y	N	N	Y	2.8	56	N	0	*RECEIVABLES_PERIOD*
22	INVENTORIESonTA	Y (+)	Y	Y	Y	Y	Y	N	Y	Y	N	N	3.3	60	Y	0	*INVENTORY_PERIOD*

(continued)

Table 4.26 *(continued)*

# Ratio	It has economic meaning (Y/N); sign of relation with PD (+,-)	Structural and Empirical Monotonicity		Means. Trimmed means and Medians are aligned with hypotheses			Histograms and Q-Q plot suggest normality of distributions Good Bad subsamples		Normality test (K-S; SW) is not significant (p-value > 0.05) Good Bad subsamples		Homogeneity of Variance (Levine's) test is not significant (p-value > 0.05)	Means difference t-test is significant (p-value < 0.05)	Means difference F-test is significant (p-value < 0.05)	F-value	AuROC value Statistical significance at 5%		Correlation > 70% Pearson's *Spearman's*	# Outliers (3 × IQR)	Conclusions
Legend	Y/N + -	Y/N	Y/N	Y/N	Y/N	Y/N	Y/N	Y/N	Y/N	Y/N	Y/N	Y/N	Y/N	Value	Value	Y/N	Ratios	Number	
23 EQUITYon PERMAN ENT_ CAPITAL	Y (-)	Y	Y	N	Y	Y	N	N	N	N	N	N	Y	9.0	60	Y		59	
24 TRADE_ PAYABLES onTL	Y (-)	Y	N	Y	Y	Y	Y	Y	Y	Y	N	Y	Y	7.9	62	Y		0	
25 DEBT_ EQUITY_ RATIO	Y (+)	N²	N	N	Y	Y	N	N	N	N	N	N	Y	6.5	49	N		116	
26 CURRENT_ RATIO	Y (-)	Y	N	Y	Y	Y	N	N	N	N	Y	N	N	0.4	55	N		48	
27 QUICK_ RATIO	Y (-)	Y	N	Y	Y	Y	N	N	Y	N	N	Y	N	N	3.5	61	Y	22	

#	Indicator	Sign															Related
28	SALESonVP	Y (–)	Y[1]	N	Y	Y	Y	N	N	N	Y	N	Y	N	0.0	55 N	169
29	SALESminus VConEBIT	Y (–)	Y	N	N	Y	Y	N	N	N	Y	N	Y	N	0.3	52 N	93
30	ROAminus IEonTL	Y (–)	Y	Y	N	Y	Y	N	N	N	N	N	Y	Y	22.0	62 Y	71 — ROI, ASSET_TURNOVER, IEonLIABILITIES, *ROEtr*, *ROI*, *ROA*
31*	EBITDAonIE	Y (–)	Y	Y	Y	Y	Y	N	N	N	Y	N	Y	N	1.7	65 Y	113
32*	EQUTYon LIABILITIES	Y (–)	Y	Y	Y	Y	Y	N	N	N	Y	N	Y	N	1.8	59 Y	63
33*	DebtEquityTr	Y (+)	Y	Y	Y	Y	Y	N	N	N	N	N	Y	Y	13.9	67 Y	101
34*	ROEtr	Y (–)	Y	N	Y	Y	Y	N	N	N	N	N	Y	Y	24.1	58 Y	109 — ROE, *ROA*, *ROAminus IEonTL*

e Variables erased from analysis after initial checks on indicators' economic meaning
* Variables added to analysis after initial checks on indicators' economic meaning
[1] Given the low probability of non-monotonicity, the indicator is considered monotonic
[2] Reciprocal value needs to be calculated
[3] Uncertain sign of the relation

significant much more frequently than mean difference, and F-test statistic is not strictly correlated with AuROC. Techniques may differ for many reasons; a basic distinction is between those depending on a mere ranking of variables' values as opposed to those depending on scale differences among variables.

To avoid excessive subjectivity when selecting the short list, it is advisable to set up a 'selection policy' to be uniformly applied to all variables, so screening indicators by a unique filter.

4.6   Estimating a model and assessing its discriminatory power

4.6.1   Steps and case study simplifications

Up until now, we have examined the steps of a statistical-based rating model development process, which are the most demanding in terms of both time and costs (data collection, cleaning, analysis sample definition, univariate and bivariate analysis, indicators transformations, short listing). These steps are crucial, since the analysis and the choices made during these steps have a huge impact on the models' final results.

Once the indicators have been short listed, the models' developers tend to:

(a) Verify the performance of multivariate models created by very different methodologies (at least logistic regression, discriminant analysis and neural networks).

(b) Test various specific choices (such as different parameters for stepwise processes used to select variables).

(c) Intervene by forcing statistical procedures (for instance, eliminating variables that enter the multivariate function with unexpected signs, including the model variables that were not selected by statistical selection procedures but which are important in order to avoid leaving some important aspects of borrower's analysis completely unexposed).

To summarize, there are a number of ways to refine the selection among short listed variables and to estimate the model in order to maximize its performance. Therefore, in this book, the development of the case study is based on a series of simplifications, which are specified below:

(a) We will only consider variables that we initially considered and/or have actually developed during examples proposed in previous paragraphs.

(b) We are going to consider all of the variables included in the dataset named W_CS_1_AnalysisSampleDataSet_7 as if they were all short listed; in other words, we will not exclude any variable from the dataset on which we performed univariate and bivariate analysis. The reasons are didactic; for instance, we will be able to check which variables have been selected through multivariate procedures and the relevance of original versus transformed variables. Additionally, we are not going to eliminate correlated variables. Occasionally, this is the approach that some model developers adopt; they perform correlation analysis only on variables that have been entered into the multivariate function and re-estimate the model once the weaker of the highly correlated pairs of variables has been eliminated from the model. This approach is useful to avoid having to calculate enormous correlation matrices when the number of variables in the long list is too large; on the other hand, this approach requires some iteration in model estimation and possible problems of convergence of estimation algorithms.

(c) We will develop two base models using linear discriminant analysis and logistic regression, and we will only test a few variants of these base models.

The simplifications listed above imply that the models we will obtain and their respective performances are purely illustrating the base results a model developer might achieve if he/she does not refine the process.

We will first focus on discriminant analysis, which will be followed by logistic regression. The former was widely used in the pioneering developments and is very useful from a didactical perspective. Nowadays, it is considered as an earlier alternative to logistic regression. The latter has two advantages: results can be directly interpreted as probabilities of default and statistical properties required for explanatory variables are less stringent than for discriminant models (independent variables do not need to be normally distributed, linearly related, or have equal within-group variances). As a result, logistic regression is preferred when data are not normally distributed or when group sizes are very unequal. However, discriminant analysis is preferred when the required assumptions are met, since it has more statistical power (less chance of 'type 2' errors, that is, accepting a false null hypothesis). Often, the two methodologies lead to very similar performance of statistical-based rating models.

4.6.2   Linear discriminant analysis

Using the 01Status variable as the target (dependent) variable, and the set of available explanatory variables, SPSS-PASW can quite easily estimate a model using linear discriminant analysis. This analysis is used to model the value of a dependent categorical variable based on its relationship to one or more predictors. From the main menu:

Analyze, Classify, Discriminant, Grouping variable: 01Status, Define range: 0 1, Continue, Independents: from ROE to ROEtr3 plus T2ROE OVERSECT430_614 SALESBin2cl1 SALESBin2cl2, Use stepwise method

We have excluded from explanatory variables T1ROE, as the transformation leading to T2ROE is preferred on a judgmental basis, and the multilevel categorical variables that have been substituted with the related dummy variables. Now, we must fill in and describe the contents of the four boxes in the upper right side of Figure 4.22.

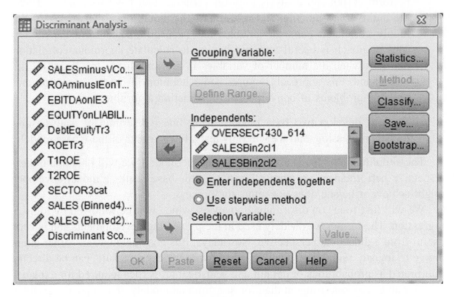

Figure 4.22   SPSS-PASW discriminant analysis window.

In *Statistics*, there are three areas. In the first, *Descriptives*, it is possible to request three analyses concerning explanatory variables: Means (and standard deviations), *Univariate ANOVAs* (test of equality of groups' means) and Box's Test (of equality of covariance matrices). Just to memorize, flag *Means*. The second area concerns *Matrices*. Do not flag any option. *Within-groups correlation* displays a pooled within-groups correlation matrix that shows the correlations between the predictors. The *Within-groups covariance* table displays a pooled within-groups covariance matrix, which may differ from the total covariance matrix. *Separate-groups covariance* displays separate covariance matrices for each group. *Total covariance* displays a covariance matrix from all cases as if they were from a single sample. The third area is relative to *Function coefficients*: select *Unstandardized*, in order to obtain the model's coefficient that can be applied to the variables' unstandardized values. In fact, discriminant function can be written either as a linear combination of standardized coefficients and variables, or as a linear combination of unstandardized coefficients and variables. By choosing this option, SPSS-PASW will not only provide standardized coefficients but also unstandardized ones. *Fisher's* classification function coefficients can be used directly for classification: a separate set of classification function coefficients is obtained for each group, and a case is assigned to the group which has the largest discriminant score (classification function value).

In *Methods* there are three options regarding stepwise procedures (to have access to these options *Use stepwise method* must be selected in the main discriminant analysis window):

(a) Again, in the area called *Method*, the indicator to be optimized at each step of the procedure can be selected. Choosing *Wilks' lambda* at each step, enter the variable that minimizes the overall Wilks' lambda. Other methods are similar. In particular, *Smallest F ratio* is a method of variable selection in stepwise analysis based on maximizing an F ratio.

(b) In the area called *Criteria*, it is possible to choose the criteria by which a variable would be entered into or would be removed from the model. If an *F-value* criterion has been chosen, a variable is entered into the model if its F-value is greater than the Entry value and is removed if the F-value is less than the Removal value. To enter more variables into the model, lower the Entry value; to remove more variables, increase the Removal value. If *probability of F* is used, a variable is entered into the model if the significance level of its F-value is less than the Entry value and is removed if the significance level is greater than the Removal value. In this case, to enter more variables into the model you must increase the Entry value whereas to remove more variables from the model you need to lower the Removal value. In the former criteria, Entry value must be higher than Removal value; in the latter, it must be the opposite. In both criteria, both values must be positive.

(c) In the area *Display*, with *Summary of steps* an output is obtained showing variables entered and removed at each step of the stepwise procedure.

In *Classify*, there are four areas of choice and the option *Replace missing values with mean*:

1. First area: *Prior Probabilities*. This option determines whether the classification is adjusted for a prior knowledge of group membership. *All groups equal* has no effect on classification, whereas choosing *Compute from group sizes* implies that the observed group sizes in your sample determine the prior probabilities of group membership and classification criteria are consequently adjusted to increase the likelihood of membership in the larger group.

2. In the second area (*Display*) there are three flags. *Casewise results*: codes for actual group, predicted group, posterior probabilities, and discriminant scores are displayed for each case. *Summary table* shows the number of cases correctly and incorrectly assigned to each of the groups based on the discriminant analysis (also called contingency table or confusion matrix). *Leave-one-out classification* is a method for out-of-sample validation of small samples for which a true subdivision of the sample into an analysis and a validation subsamples would be inconvenient. Each case in the

analysis is classified by the functions derived from all cases other than that case. Other similar methods are the Jacknife and the Bootstrapping methods.

3. In the third area, *Use Covariance Matrix*, you can choose different methods of classification of cases (using a within-groups covariance matrix or a separate-groups covariance matrix).

4. In the last area, *denominated Plots*, there are three alternatives. *Combined-groups* create a histogram. *Separate-groups* create histograms for the two groups. *Territorial map* plots the boundaries used to classify cases into groups based on function values. It is not displayed if there is only one discriminant function as it occurs in our application (having a dichotomous variable as dependent, we have only one discriminant function).

In *Save* there are three options concerning which output variables to add to the existing dataset: *predicted group membership* specifies the estimated group to which a given observation belongs to, according to the model; *discriminant scores* saves the score values of each borrower; *probabilities of group membership* transforms scores into probabilities of belonging to the performing or the default groups (in this case, two variables are added).

We select:

Statistics, Means, Univariate ANOVAs, Box's M, Unstandardized, Continue,

Method, Wilks' lambda, Use probability of F (using standard values proposed by SPSS-PASW), Summary of steps, Continue

Classify, All groups equal, Within-groups, Summary table, Leave-one-out classification, Replace missing values with mean, Continue

Save, Discriminant scores, Continue

By choosing *Paste* and *Run* we get syntax, output, and new variables in the dataset. Save files as:

W_AnalysisSampleSyntax_9MDA

W_AnalysisSampleOutput_9MDA

W_CS_1_AnalysisSampleDataSet_8MDA

Now consider the output file *W_AnalysisSampleOutput_9MDA*. First of all, there is a summary of the cases used; there are 1272 valid observations, no missing values or out-of-range group codes. Then, in the table *Group Statistics*, we can read the number of valid cases, the mean and the standard deviation for each variable, for each group (performing, default) and for the sample as a whole.

Table 4.27 reports an extract of tests of equality of group means, showing two statistics that are practically equivalent.

Table 4.27   Tests of equality of group means.

Tests of equality of group means					
	Wilks' Lambda	F	df1	df2	Sig.
NetProfit/Equity	0.988	14.843	1	1265	0.000
EBITDA/Sales	1.000	0.001	1	1265	0.972
EBIT/OperatingAssets	0.983	21.827	1	1265	0.000
CurrentIncome/TotalAssets	0.983	22.152	1	1265	0.000
EBITDA/ValueOfProduction	1.000	0.015	1	1265	0.902
EBIT/Sales	1.000	0.006	1	1265	0.940
Sales/TotalAssets	0.982	22.926	1	1265	0.000
Sales/Inventories	1.000	0.353	1	1265	0.552
Sales/TradeReceivables	1.000	0.324	1	1265	0.569
TradeReceivables/Sales/360	1.000	0.011	1	1265	0.917
Inventories/Sales/360	1.000	0.068	1	1265	0.794
TradePayables/Purchases/360	1.000	0.357	1	1265	0.550
(TradeReceivables+Inventories- TradePayables)/Sales/360	1.000	0.026	1	1265	0.871
InterestExpenses/EBITDA	1.000	0.548	1	1265	0.459
NetInterestExpenses/EBITDA	1.000	0.551	1	1265	0.458
InterestExpenses/Liabilities	0.982	23.751	1	1265	0.000
InterestExpenses/FinancialDebts	1.000	0.503	1	1265	0.478
GrossOfTaxProfit/IncomeOnOrd.Act.	1.000	0.385	1	1265	0.535
Taxes/GrossOfTaxProfit	0.999	0.974	1	1265	0.324
Intangibles/TotalAssets	1.000	0.013	1	1265	0.910
TradeReceivables/TotalAssets	0.998	2.836	1	1265	0.092
Inventories/TotalAssets	0.997	3.396	1	1265	0.066
Equity/PermanentCapital	0.993	8.935	1	1265	0.003
TradePayables/TotalLiabilities	0.994	7.840	1	1265	0.005
Debt/Equity	0.995	6.775	1	1265	0.009
CurrentAssets/CurrentLiabilities	1.000	0.358	1	1265	0.550
(CurrentAssets-Inventories)/CurrentLiabilities	0.997	3.544	1	1265	0.060
Sales/ValueOfProduction	1.000	0.027	1	1265	0.870
(Sales-VariableCosts)/EBIT	1.000	0.312	1	1265	0.577
ROA-(InterestExpenses/TotalLiabilities)	0.983	21.919	1	1265	0.000
EBITDAonIE	0.999	1.702	1	1265	0.192

F-ratio is between groups' variance divided by within groups' variance, namely, Explained variance/Unexplained variance (or Sum of Squares between group's means/Sum of Squares within groups means). The higher its value, the lower is the probability that group means are equal. The p-value of the relative test (*Sig.* in the last column) tests the null hypothesis that there is no difference between the group means. We are able to reject Ho if we get an F-test statistic high enough to have a p-value lower than the set alpha level of significance (usually 5%).

Wilks' Lambda is the ratio of the sum of squares in the groups and the sum of overall squares, namely, Unexplained variance/Total variance. The ratio tends to one when group means are similar, whereas it tends to zero when the variance in groups is small compared to the overall variance.

Variables that are able to minimize Wilks' Lambda are those which maximize the F-ratio; p-values in the last column are referred to both measures.

To select variables to include in the model, Wilks' Lambda was chosen. This method, at each step, selects the variable (to enter into the model) that minimizes the overall Wilks' Lambda and the corresponding p-value (it is equivalent to maximizing F-value): the variable is entered into the model if the level of significance of F is lower than 5%; it is removed from the model if this value is beyond 10%.

Table 4.28 shows a synthesis of the stepwise process, outlining the seven variables entered into the model, step by step. Table 4.29 shows an extract of the (very long) table of variables that, at each step, are not included into the model; note that all variables not included present a higher Wilks' Lambda statistic. The process ends once there are no other variables that can be entered or removed from the model which would improve it.

Table 4.28   Stepwise statistics.

Variables entered/removed[a,b,c,d]								
Step		Wilks' Lambda						
						Exact F		
	Entered	Statistic	df1	df2	df3	Statistic	df1	df2	Sig.
1	ROETr	.981	1	1	1265,000	24.075	1	1,265.000	.000
2	InterestExpenses/ Liabilities	.963	2	1	1265,000	24.369	2	1,264.000	.000
3	INVENTORY_ PERIOD3	.955	3	1	1265,000	19.854	3	1,263.000	.000
4	Equity/Permanent Capital	.947	4	1	1265,000	17.520	4	1,262.000	.000
5	EQUITYon PERMANENT_ CAPITAL3	.935	5	1	1265,000	17.622	5	1,261.000	.000
6	SALESBin2cl1	.931	6	1	1265,000	15.573	6	1,260.000	.000
7	ROA-(InterestExpenses/ TotalLiabilities)	.928	7	1	1265,000	13.991	7	1,259.000	.000

At each step, the variable that minimizes the overall Wilks' Lambda is entered.
[a]Maximum number of steps is 146.
[b]Maximum significance of F to enter is .05.
[c]Minimum significance of F to remove is .10.
[d]F level, tolerance, or VIN insufficient for further computation.

Table 4.29   Variables not entered into the model at each step of the stepwise procedure (extract from SPSS-PASW overall table).

Step		Variables not in the analysis			
		Tolerance	Min. Tolerance	Sig. of F to Enter	Wilks' Lambda
0	NetProfit/Equity	1.000	1.000	0.000	0.988
	EBITDA/Sales	1.000	1.000	0.972	1.000
	EBIT/OperatingAssets	1.000	1.000	0.000	0.983
	CurrentIncome/TotalAssets	1.000	1.000	0.000	0.983
	EBITDA/ValueOfProduction	1.000	1.000	0.902	1.000
	EBIT/Sales	1.000	1.000	0.940	1.000
	Sales/TotalAssets	1.000	1.000	0.000	0.982
	Sales/Inventories	1.000	1.000	0.552	1.000
	Sales/TradeReceivables	1.000	1.000	0.569	1.000
	TradeReceivables/Sales/360	1.000	1.000	0.917	1.000
	Inventories/Sales/360	1.000	1.000	0.794	1.000
	TradePayables/Purchases/360	1.000	1.000	0.550	1.000
	(TradeReceivables+Inventories-TradePayables)/Sales/360	1.000	1.000	0.871	1.000
	InterestExpenses/EBITDA	1.000	1.000	0.459	1.000
	NetInterestExpenses/EBITDA	1.000	1.000	0.458	1.000
	InterestExpenses/Liabilities	1.000	1.000	0.000	0.982
	InterestExpenses/FinancialDebts	1.000	1.000	0.478	1.000
	GrossOfTaxProfit/IncomeOnOrd.Act.	1.000	1.000	0.535	1.000
	Taxes/GrossOfTaxProfit	1.000	1.000	0.324	0.999
	Intangibles/TotalAssets	1.000	1.000	0.910	1.000
	TradeReceivables/TotalAssets	1.000	1.000	0.092	0.998
	Inventories/TotalAssets	1.000	1.000	0.066	0.997
	Equity/PermanentCapital	1.000	1.000	0.003	0.993
	TradePayables/TotalLiabilities	1.000	1.000	0.005	0.994
	Debt/Equity	1.000	1.000	0.009	0.995

Table 4.30 gives more details on each step of the procedure. Note that there are no variables removed from the model after they have been entered; in fact, none has a level of significance of F greater than the set level of removal (10%). Note also that Wilks' Lambdas' and p-values for a given variable change at each step, because they depend on the contribution to explain variability offered by other variables in the model.

Table 4.31 reports the model's Wilks' Lambda statistic. In effect, Wilks' Lambda is a statistic used in multivariate analysis of variance (MANOVA) to test whether there are differences between the means of identified groups of subjects on a combination of dependent variables. Wilks' Lambda performs, in a

Table 4.30   Variables entered into the model at each step of stepwise procedure.

Step	Variables in the analysis	Tolerance	Sig. of F to remove	Wilks' Lambda
1	ROETr	1.000	0.000	
2	ROETr	1.000	0.000	0.982
	InterestExpenses/Liabilities	1.000	0.000	0.981
3	ROETr	0.999	0.000	0.973
	InterestExpenses/Liabilities	0.997	0.000	0.974
	INVENTORY_PERIOD3	0.997	0.001	0.963
4	ROETr	0.999	0.000	0.965
	InterestExpenses/Liabilities	0.996	0.000	0.967
	INVENTORY_PERIOD3	0.996	0.001	0.955
	Equity/PermanentCapital	0.999	0.002	0.955
5	ROETr	0.999	0.000	0.952
	InterestExpenses/Liabilities	0.996	0.000	0.953
	INVENTORY_PERIOD3	0.996	0.001	0.942
	Equity/PermanentCapital	0.717	0.000	0.952
	EQUITYonPERMANENT_CAPITAL3	0.717	0.000	0.947
6	ROETr	0.999	0.000	0.948
	InterestExpenses/Liabilities	0.988	0.000	0.951
	INVENTORY_PERIOD3	0.996	0.002	0.938
	Equity/PermanentCapital	0.717	0.000	0.949
	EQUITYonPERMANENT_CAPITAL3	0.715	0.000	0.943
	SALESBin2cl1	0.989	0.025	0.935
7	ROETr	0.998	0.000	0.945
	InterestExpenses/Liabilities	0.463	0.045	0.931
	INVENTORY_PERIOD3	0.996	0.001	0.935
	Equity/PermanentCapital	0.716	0.000	0.946
	EQUITYonPERMANENT_CAPITAL3	0.710	0.000	0.941
	SALESBin2cl1	0.989	0.025	0.932
	ROA-(InterestExpenses/TotalLiabilities)	0.464	0.039	0.931

multivariate setting – with a combination of dependent variables, the same role as the F-test performs in one-way analysis of variance. Wilks' Lambda is a direct measure of the proportion of variance in the combination of dependent variables that is unaccounted for by the independent variable (the grouping variable or factor). Wilks' Lambda statistic can be transformed (mathematically adjusted) to a

Table 4.31   Wilks' Lambda.

						Exact F			
Step	Number of variables	Lambda	df1	df2	df3	Statistic	df1	df2	Sig.
1	1	0.981	1	1	1265	24.075	1	1,265.000	0.000
2	2	0.963	2	1	1265	24.369	2	1,264.000	0.000
3	3	0.955	3	1	1265	19.854	3	1,263.000	0.000
4	4	0.947	4	1	1265	17.520	4	1,262.000	0.000
5	5	0.935	5	1	1265	17.622	5	1,261.000	0.000
6	6	0.931	6	1	1265	15.573	6	1,260.000	0.000
7	7	0.928	7	1	1265	13.991	7	1,259.000	0.000

statistic which has an approximate F distribution. This makes it easier to calculate the p-value.

Step by step, Wilks' Lambda values start decreasing from one (indicating an improvement of the model); the related F statistic p-values are always close to zero.

Box's M Test concerns one of the statistical requirements of discriminant analysis: equal variance and covariance matrices of groups. The lower the value of M (and the higher the p-value beyond 5%), the more probable the equality of variance and covariance matrices. Table 4.32 shows an initial weakness in the model we are developing in this case study.

Table 4.32   Box's M test results.

Test results		
Box's M		9903.486
F	Approx.	335.890
	df1	28
	df2	25 526.682
	Sig.	0.000

Tests null hypothesis of equal population covariance matrices.

In Table 4.33, the eigenvalues table provides information regarding the relative efficacy of each discriminant function. Eigenvalues are linked with the percentage of variance among groups explained by each discriminant function in respect to the total variance explained by all discriminant functions. In our case study, the grouping variable has only two levels, so we have only one discriminant function

Table 4.33   Summary of canonical discriminant functions.

Eigenvalues			
Function	Eigenvalue	% of variance	Cumulative %	Canonical correlation
1	0.078[a]	100.0	100.0	0.269

Wilks' Lambda			
Test of Function(s)	Wilks' Lambda	Chi-square	df	Sig.
1	0.928	94.504	7	0.000

[a]First 1 canonical discriminant functions were used in the analysis.

and it accounts for 100% of the total explained variance. When there are two groups, the canonical correlation (Eta) is the most useful measure in the table; it is the equivalent to Pearson's correlation between the discriminant scores and the groups. Instead, Wilks' Lambda indicates the proportion of the total sum of squares not explained by the differences in the two groups. The two statistics are mathematically related by:

$$\Lambda_i - \eta_i^2 = 1$$

and they measure the same phenomenon.

The model estimated using discriminant analysis is represented in Table 4.34, showing canonical coefficients of the seven variables selected by the stepwise procedure and the constant term. The algebraic sum of the products of these coefficients with their respective variables gives the 'score' of a given borrower.

Table 4.34   Canonical discriminant function coefficients.

Canonical discriminant function coefficients	
	Function
	1
InterestExpenses/Liabilities	0.080
Equity/PermanentCapital	0.002
ROA-(InterestExpenses/TotalLiabilities)	0.001
ROETr	0.000
INVENTORY_PERIOD3	0.004
EQUITYonPERMANENT_CAPITAL3	−0.008
SALESBin2cl1	−0.512
(Constant)	−0.167

Unstandardized coefficients

So, the complete expression of the resulting discriminant function is (only the first three decimals for each coefficient are shown):

$$\text{Score} = -0,163 + 0,080 \times \text{InterestExpenses/Liabilities} + 0,002 \times \text{Equity/}$$

$$\text{PermanentCapital} + 0,001 \times [\text{ROA} - (\text{InterestExpenses/TotalLiabilities})]$$

$$+ 0,000 \times \text{ROETr} + 0,004 \times \text{Inventory\_Period3} - 0,008$$

$$\times \text{EquityOnPermanent\_Capital3} - 0,515 \times \text{SalesBin2cl1}$$

We are now in a position to rank borrowers using scores. To determine whether a higher score indicates higher default risk or creditworthiness, it is necessary to examine Table 4.35, which reports so called 'centroids'. They are the means of the scores of the two subsamples (performing and defaulted borrowers). As the score mean for the subsample of defaulted borrowers is greater than that of performing borrowers, we deduce that higher scores indicate higher default risk.

Table 4.35   Functions at groups centroids.

Functions at group centroids	
01STATUS	Function
	1
performing	−0.057
default	1.361

Unstandardized canonical discriminant functions evaluated at group means

A key question regards the possibility of interpreting the level of coefficients as the contribution of variables to the score. The answer is no for two reasons:

The first reason is that different variables have different range of variability; so, the coefficient is estimated also taking into account this aspect and cannot be interpreted as the 'importance' of the variable in the model. This issue can be overcome by calculating standardized values for each variable. Calculating the value of the variable minus the sample mean, and then dividing by the sample standard deviation, allows a distribution where the mean is zero and the standard deviation is one. In this case, all variables have the same scale.

Of course, if we use standardized variables, we need to combine them with standardized coefficients. The upper left side of Table 4.36 presents canonical discriminant function coefficients to be multiplied by standardized values of variables. Note that the constant term has disappeared and that coefficient signs remain the same.

The second reason is that a given variable's coefficient depends on other variables included into the model; as there is some correlation among variables, the estimation of a coefficient takes into account co-movements in other variables. This

Table 4.36   Standardized canonical discriminant function coefficients and structure matrix.

Standardized canonical discriminant function coefficients	
	Function
	1
InterestExpenses/Liabilities	0.309
Equity/PermanentCapital	0.609
ROA-(InterestExpenses/TotalLiabilities)	0.318
ROETr	−0.501
INVENTORY_PERIOD3	0.334
EQUITYonPERMANENT_CAPITAL3	−0.517
SALESBin2cl1	−0.236

Structure matrix			
	Function		Function
	1		1
ROETr	−0.495	ROAminuslEonTL3[a]	−0.156
InterestExpenses/Liabilities	0.491	INVENTORY_TURNOVER3[a]	−0.147
Sales/TotalAssets[a]	0.481	SALESBin2cl2[a]	0.145
CurrentIncome/TotalAssets[a]	0.474	COMMERCIAL_WC_	0.128
ROA-(InterestExpenses/	0.472	PERIOD3[a]	
TotalLiabilities)		EQUITYonLIABILITIES3[a]	−0.126
EBIT/OperatingAssets[a]	0.472	ROI3[a]	−0.119
DebtEquityTr[a]	0.359	EBITDAonVP3[a]	−0.115
NetProfit/Equity[a]	0.346	ROA3[a]	−0.115
INVENTORY_PERIOD3	0.317	(CurrentAssets-Inventories)/	−0.110
		CurrentLiabilities[a]	
Equity/PermanentCapital	0.301	EBITDAonIE[a]	−0.110
EBITDAonIE3[a]	−0.244	T2ROE[a]	0.107
IEonLIABLITIES3[a]	0.244	TRADE_PAYABLESonTL3[a]	−0.105
INVENTORIESonTA3[a]	0.227	TradePayables/TotalLiabilities[a]	−0.105
Inventories/TotalAssets[a]	0.227	Debt/Equity[a]	−0.105
EQUITYonPERMANENT_	−0.218	ROETr3[a]	−0.102
CAPITAL3			
SALESBin2cl1	−0.206	ASSETS_TURNOVER3[a]	−0.099
DebtEquityTr3[a]	0.206	EQUITYonLIABILITIES[a]	−0.088
QUICK_RATIO3[a]	−0.172	ROS3[a]	−0.087
IEonEBITDA3[a]	0.170	EBITDAonSALES3[a]	−0.086
NIEonEBITDA3[a]	0.170	Sales/Inventories[a]	−0.070
TradeReceivables/TotalAssets[a]	−0.164	SALESonVP3[a]	−0.066
TRADE_RECEIVABLES onTA3[a]	−0.164		

Table 4.36   *(continued)*

Structure matrix			
	Function		Function
	1		1
EXTR_I_C_onCURRENT_INCOME3[a]	−0.062	EBITDA/Sales[a]	0.024
RECEIVABLES_TURNOVER3[a]	0.062	EBIT/Sales[a]	0.023
RECEIVABLES_PERIOD3[a]	−0.048	ROE3[a]	−0.018
CurrentAssets/CurrentLiabilities[a]	0.043	Intangibles/TotalAssets[a]	0.018
DEBT_EQUITY_RATIO3[a]	0.039	INTANGIBLESonTA3[a]	0.018
PAYABLES_PERIOD3[a]	0.038	EBITDA/ValueOfProduction[a]	0.013
SALESminusVConEBIT3[a]	0.036	IEonFINANCIAL_DEBTS3[a]	0.012
Taxes/GrossOfTaxProfit[a]	0.034	(Sales-VariableCosts)/EBIT[a]	0.011
InterestExpenses/FinancialDebts[a]	−0.034	GrossOfTaxProfit/IncomeOnOrd.Act.[a]	−0.010
Sales/TradeReceivables[a]	0.029	InterestExpenses/EBITDA[a]	−0.010
Inventories/Sales/360[a]	0.029	NetInterestExpenses/EBITDA[a]	−0.010
CURRENT_RATIO3[a]	0.029	TradeReceivables/Sales/360[a]	−0.003
(TradeReceivables+Inventories-TradePayables)/Sales/360[a]	0.029	Sales/ValueOfProduction[a]	−0.002
ROE (Binned)[a]	−0.029	TradePayables/Purchases/360[a]	−0.002
		OVERSECT430_614[a]	0.001

Pooled within-groups correlations between discriminating variables and standardized canonical discriminant functions
Variables ordered by absolute size of correlation within function.
[a]This variable not used in the analysis.

is why the contribution of a given variable to the score (its 'importance' in the model) is better expressed by the Structure Matrix, which reports the correlation coefficient between scores and a given variable's values: the higher the coefficient, the stronger the relationship between variable's values and scores, and the higher the variable's contribution to the discriminant function. The Structure Matrix lists indicators in a descending order of importance (Table 4.36).

For instance, take ROETr; first of all, note the difference between the standardized canonical function coefficient and the coefficient value in the Structure Matrix. Then, compare it with Equity/PermanentCapital coefficients (the standardized canonical function coefficient is greater for the latter ratio, whereas in the

Structure Matrix the latter ratio has a lower coefficient than ROETr). In addition, it is apparent that many indicators which have a high correlation with scores and are listed quite high in the Structure Matrix have not been entered into the discriminant function. Evidently, they are correlated with other ratios in the model and their marginal contribution to the model would be lower than that of other indicators and, when individually considered, appear to be less relevant.

Now we can comment on some aspects concerning the structure and the robustness of the estimated model. As we did not actually select a short list of indicators compliant with statistical requirements assessed by univariate and bivariate analyses, it would be useful to discuss the characteristics of variables entered into the model. Then, we need to check coefficients' signs, the significance of individual coefficients and the coverage of relevant information categories. Lastly, we have to evaluate the overall performance of the model.

We are not going to comment on all variables in the model, thereby leaving this duty to the reader. Consider, for instance, InterestExpenses/Liabilities: this indicator has one of the highest AuROC, the F-test states that the means difference is significant, central tendency measures (mean, trimmed mean, and median) are aligned with working hypothesis, it is (both endogenously and empirically) monotonically related with default risk, and has a low number of outliers. However, it is not normally distributed, there is no homogeneity of variance, and an independent t-test does not confirm that means are statistically significant. The sign of its coefficient in the discriminant function is coherent with expectations, and the sign remains stable in the structure matrix. There is a (Pearson) correlation of 73% with the indicator of spread between ROA and InterestExpenses/TotalLiabilities, statistically significant at 1%.

Considering the main issues in the model structure, it is important to outline:

(a) The indicator Equity/PermanentCapital is entered into the model with two different configurations: the original variable (with a coefficient sign coherent with expectations) and the transformed variable confining outliers (59 cases) to three times the interquartile range from the interquartile range boundaries (with an unexpected coefficient sign). Coefficient signs in the Structure Matrix remain stable. Because, apart from the tails, the distributions of the two indicators are the same, they must be strongly correlated. To verify this presumption, request a correlation analysis of variables entered into the model (with *Analyze, Correlate, Bivariate*, choosing the six scale variables in the model, *Paste, Run selection*, and then *Save* for. sps and. spv files; results are in Table 4.37). We see that Pearson's correlation is 52% (so, it is lower than the danger level of 70% we have chosen), but Spearman's correlation is practically 100%.

(b) The indicator of spread between ROA and InterestExpenses/TotalLiabilities enters into the model with a positive sign. It is therefore incoherent with expectations, given that the observation of centroids stated that the score is positively related with default risk. In addition, the variable is strongly

Table 4.37 Pearson's and Spearman's correlations.

Correlations

		Interest Expenses/ Liabilities	Equity/ Permanent Capital	ROA- (InterestExpenses/ TotalLiabilities)	ROETr	INVENTORY_ PERIOD3	EQUITYon PERMANENT_ CAPITAL3
InterestExpenses/ Liabilities	Pearson Correlation	1	-0.013	0.734**	0.001	-0.041	-0.034
	Sig. (2-tailed)		0.638	0.000	0.983	0.145	0.221
	N		1272	1272	1272	1272	1272
Equity/Permanent Capital	Pearson Correlation	-0.013	1	0.002	0.005	-0.009	0.524**
	Sig. (2-tailed)	0.638		0.955	0.868	0.738	0.000
	N	1272	1272	1272	1272	1272	1272
ROA-(Interest Expenses/ TotalLiabilities)	Pearson Correlation	0.734**	0.002	1	0.008	-0.040	0.028
	Sig. (2-tailed)	0.000	0.955		0.765	0.159	0.310
	N	1272	1272	1272	1272	1272	1272
ROETr	Pearson Correlation	0.001	0.005	0.008	1	-0.035	0.023
	Sig. (2-tailed)	0.983	0.868	0.765		0.217	0.413
	N	1272	1272	1272	1272	1272	1272
INVENTORY_PERIOD3	Pearson Correlation	-0.041	-0.009	-0.040	-0.035	1	-0.030
	Sig. (2-tailed)	0.145	0.738	0.159	0.217		0.290
	N	1272	1272	1272	1272	1272	1272
EQUITYon PERMANENT_ CAPITAL3	Pearson Correlation	-0.034	0.524**	0.028	0.023	-0.030	1
	Sig. (2-tailed)	0.221	0.000	00.310	0.413	0.290	
	N	1272	1272	1272	1272	1272	1272

(continued)

Table 4.37 (continued)

Correlations

		Interest Expenses/ Liabilities	Equity/ Permanent Capital	ROA- (InterestExpenses/ TotalLiabilities)	ROETr	INVENTORY_ PERIOD3	EQUITYon PERMANENT_ CAPITAL3
Spearman's rho	Interest Expenses/ Liabilities						
	Correlation Coefficient	1.000	-0.167**	-0.124**	0.005	0.040	-0.167**
	Sig. (2-tailed)	.	0.000	0.000	0.852	0.154	0.000
	N	1272	1272	1272	1272	1272	1272
	Equity/ Permanent Capital						
	Correlation Coefficient	-0.167**	1.000	0.136**	-0.010	-0.062*	1.000**
	Sig. (2-tailed)	0.000	.	0.000	0.730	0.026	0.000
	N	1272	1272	1272	1272	1272	1272
	ROA-(Interest Expenses/ TotalLiabilities)						
	Correlation Coefficient	-0.124**	0.136**	1.000	0.781**	-0.338**	0.136**
	Sig. (2-tailed)	0.000	0.000	.	0.000	0.000	0.000
	N	1272	1272	1272	1272	1272	1272
	ROETr						
	Correlation Coefficient	.005	-0.010	0.781**	1.000	-0.238**	-0.010
	Sig. (2-tailed)	0.852	0.730	0.000	.	0.000	0.733
	N	1272	1272	1272	1272	1272	1272
	INVENTORY_ PERIOD3						
	Correlation Coefficient	0.040	-0.062*	-0.338**	-0.238**	1.000	-0.062*
	Sig. (2-tailed)	0.154	0.026	0.000	0.000	.	0.026
	N	1272	1272	1272	1272	1272	1272
	EQUITYon PERMANENT_ CAPITAL3						
	Correlation Coefficient	-0.167**	1.000**	0.136**	-0.010	-0.062*	1.000
	Sig. (2-tailed)	0.000	0.000	0.000	0.733	0.026	.
	N	1272	1272	1272	1272	1272	1272

**Correlation is significant at the 0.01 level (2-tailed).
*Correlation is significant at the 0.05 level (2-tailed).

correlated with both InterestExpenses/Liabilities (Pearson's correlation) and ROETr (Spearman's correlation). The number of outliers is 71.

All other variables in the model have coefficient signs coherent with expectations. Moreover, the negative sign of the coefficient of the only dummy variable entered into the model is correct, because the first dimensional class has a lower default rate than other classes.

In addition, all variables show signs of canonical coefficients identical to those of coefficients in the Structure Matrix; if there were differences of signs for the same variable it would indicate strong multicolinearity problems among variables. If signs were different, we would also have troubles in the signs' coherence with working hypotheses, as one of the two coefficients would be unexpected.

Above all, all variables are endogenously monotonic and, apart from ROETr, they also are all empirically monotonic.

Eventually, the seven selected variables can be considered well distributed among the different perspectives of a typical credit analysis, covering profitability, debt burden, financial structure, working capital, and size.

Examine the model's overall performance. The simplest approach to evaluate a model's discriminatory capability is based on the contingency table, where actual good and bad borrowers are cross-referenced with the model prediction in terms of good and bad. It allows us to check the percentage of correct reclassifications, as well as type 1 and type 2 error rates (respectively, predicting a bad borrower as good, and a good borrower as bad).

In Table 4.38, below the in-sample classification results, leave-one-out procedure results have been reported. The model performance does not deteriorate much in the automatic out-of-sample validation. Nevertheless, a true out-of-sample validation is preferable, and possibly out-of-time and out-of-universe ones (that is,

Table 4.38   Classification results[b,c].

		01STATUS	Predicted group membership		Total
			performing	default	
Original	Count	performing	1081	140	1221
		default	35	16	51
	%	performing	88.5	11.5	100.0
		default	68.6	31.4	100.0
Cross-validated[a]	Count	performing	1075	146	1221
		default	36	15	51
	%	performing	88.0	12.0	100.0
		default	70.6	29.4	100.0

[a]Cross validation is done only for those cases in the analysis. In cross validation. each case is classified by the functions derived from all cases other than that case.
[b]86.2% of original grouped cases correctly classified.
[c]85.7% of cross-validated grouped cases correctly classified.

using samples coming from different years and different populations from those used to develop the model).

The contingency matrix in Table 4.38 shows that the model has poor performance, with about 70% type 1 error rate: missing to detect 70% of would-be defaulters would be very costly for the bank.

In any case, a contingency table simply demonstrates the capability of the model in separating borrowers into two categories: expected good and expected bad borrowers. Nothing is said on the level of risk, both based on an ordinal scale (ratings) and on cardinal numbers (default risk).

In addition, results strictly depend on the cut-off of scores chosen to separate good and bad predicted borrowers. The cut-off score is usually automatically set as the average of score means in the two subsamples (that is to say, of the two centroids) when prior probabilities (to belong to one of the two subgroups) are set equally. This was our choice, and in the SPSS-PASW output this is clearly stated (Table 4.39).

Table 4.39   Prior probabilities for groups.

Prior probabilities for groups			
01STATUS	Prior	Cases used in analysis	
		Unweighted	Weighted
performing	0.500	1216	1216
default	0.500	51	51
Total	1.000	1267	1267

AuROC is a more general performance indicator. In practice, it summarizes the model's performance for each possible chosen cut-off. In fact, each point of the curve in Figure 4.23 represents the classificatory performance we would obtain for a given cut-off. The cut-off level is not shown in the graph; instead, a ROC curve shows for each undisclosed cut-off:

- On the X-axis, the classification error rate of actually good borrowers (false positive, FP); it is a type 2 error. The complement to one represents the correct classification rate of actually good borrowers (true negative, TN).

- On the Y-axis, the correct classification rate of actually bad borrowers (true positive, TP). In this case, the complement to one represents the classification error rate of actually bad borrowers (false negative, FN); it is a type 1 error.

Therefore, a ROC curve is a graphical representation of the predictive capability of a model for each level of scores cut-offs. It is also useful to realize if a model has better performance than other models in some score intervals and worse in other score intervals. From this curve, a summary indicator of model performance can be calculated: AuROC –representing the area under ROC curve.

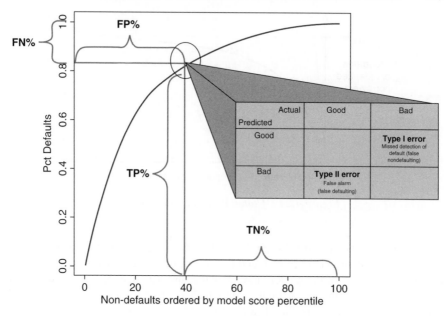

Figure 4.23   ROC curve and contingency tables.

A similar graphical representation is offered by CAP (Cumulative Accuracy Profile), also known as Lorenz curves (Figure 4.24); the synthetic associated performance indicator is called Gini ratio, or Accuracy Ratio (AR). To build a CAP, as for ROC curves, it is first of all necessary to sort borrowers from the worst score to the best score in terms of risk. Then, for CAP, it is indicated:

- on the X-axis, the cumulative percentage of borrowers (both performing and defaulted),

- on the Y-axis the cumulative percentage of bad borrowers.

presenting a score value not greater than the selected cut-off. So, the Y-axis has exactly the same meaning we stated for ROC curves, whereas the X-axis considers all borrowers and not only non-defaulted ones.

For CAP, if the system is able to perfectly discriminate between good and bad borrowers, the curve obtained considering all possible cut-off values reaches 100% of defaults on the Y-axis when, on the X-axis, it has reached a percentage of borrowers exactly equal to the default rate of the sample in use. If the model is completely naïve, the curve would be a straight line lying on the diagonal line of the graph. A real model's curve will be somewhere in between the naïve model's curve and the perfect model's curve; the closer the curve is to the perfect model's curve (and the further away it is from the naïve model's curve) the better the model will be.

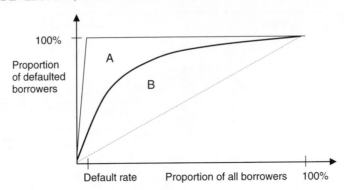

Figure 4.24   Cumulative accuracy profile.

The ratio of the area B between the CAP of the model being validated and the CAP of the random (naïve) model, and the area A between the CAP of the perfect model and the CAP of the random model, is the summary statistic called Accuracy Ratio (or Gini coefficient). A high value which is close to one indicates that the evaluated model is close to the ideal model.

In addition, judgment-based ratings, such as those offered by rating agencies, can be evaluated using Gini ratio and AuROC. Furthermore, classification performance can be measured on various time horizons. The usual one-year time frame leads to Figure 4.25, where the mean of one-year Ratings Performances between 1981 and 2008 for corporate issuers rated by Standard & Poor's is shown (Standard & Poor's, 2009a). Figure 4.26 is based on a five-year distance of the observation period from time zero (time of rating assignment). Of course, the discriminatory power diminishes when a longer time horizon is considered because it is more difficult to predict what will happen in the far future than in the near future. However, performance remains satisfactory.

Note that specific cut-offs have been indicated on the curves; these correspond to agencies' rating classes. So, for each rating class, as we did for each cut-off score, it is possible to identify the percentage of correct (TP and TN) and wrongful (FP and FN) classifications, that is to say, contingency table data.

Between AR and AuROC there is a mathematical relation we can use to transform AR into AuROC and vice versa:

$$AR = 2(AuROC - 0.5)$$

AuROC values go from a minimum of 50% to a maximum of 100%, whereas AR values move between 0% and 100%. Figure 4.27 shows the relationship between these measures.

Table 4.40 reports a synthesis of Standard & Poor's corporate ratings performances. In the long run (1981–2008), performance in the non-financial sector is superior to that of the financial sector on every time horizon considered, while, for both sectors, performance naturally decreases as the time horizon becomes greater.

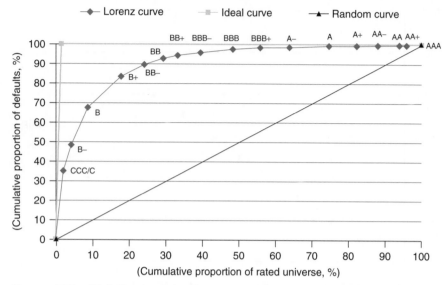

Figure 4.25 Global one-year relative corporate ratings performance (1981–2008).

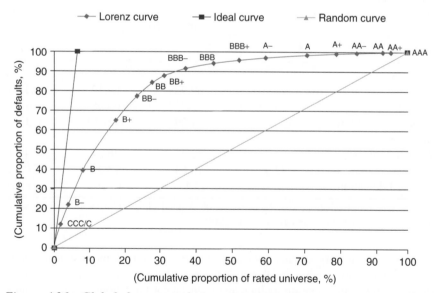

Figure 4.26 Global five-year relative corporate ratings performance (1981–2008).

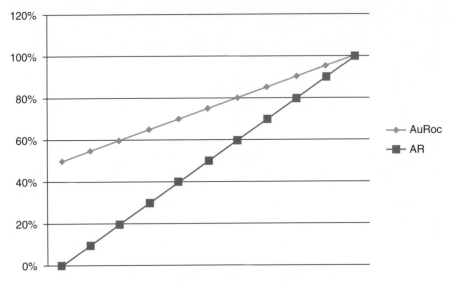

Figure 4.27   Corresponding values of AuROC and accuracy ratio.

Table 4.40   Gini Coefficients for Global Corporates By Broad Sector (1981–2008).

Sector	–Time horizon (years)–			
	1	3	5	7
Financial				
Weighted average	77.76	69.92	64.12	61.88
Average	78.53	72.43	66.08	61.81
Standard deviation	(23.94)	(14.09)	(13.80)	(11.00)
Nonfinancial				
Weighted average	84.32	78.87	76.24	74.47
Average	82.95	76.64	73.20	70.22
Standard deviation	(6.51)	(5.11)	(5.16)	(4.98)

Source: Default, Transition, and Recovery: 2008 Annual Global Corporate Default Study And Rating Transitions. Standard & Poor's, 2009.

Public availability of this data may be seen as very important in order to have a benchmark to appraise the classification power of banks' internal ratings. However, there are some difficulties due to technical limitations of the measures. Firstly, when the number of cut-offs is limited, as in the case of judgment-based rating, curves are obviously obliged to interpolate different points. When there are a different number of interpolation points, resulting performance measures cannot be perfectly comparable. Secondly, these measures are 'sample dependent', so they can only be used correctly to compare models' performance calculated on the same sample. In the research carried out by Moody's (Sobehart, Keenan, and Stein, 2000), different quantitative models were compared on the same out-of sample and out-of-time sample; their performance in terms of ARs range between 50 and 75% (that is between 75 and 87.5% of AuROC). In general, for banks, as well as for national supervisory institutions, it is extremely difficult to compare different models' performances precisely, because there is no unique reference dataset and because it is complex and burdensome to build it. The reason is that many indicators, above all those coming from qualitative and behavioral data, are so specific in different banks that they are not replicable on a common reference dataset. Therefore, model comparisons in terms of AR and AuROC are typically a large approximation. In addition, it is crucial that at least the time frame (of data, times zero, and observation periods) and the target market segments are the same. Eventually, the datasets used to calculate AR and AuROC measures should be similarly in-the-sample or out-of-sample and/or out-of-time and/or out-of-universe for each model.

How can we assess performance of the model we estimated? SPSS-PASW has a simple procedure for drawing ROC curves and calculating AuROC. From the main menu:

Analyze, RocCurve, Test variable: Discriminant Score (it is the new variable added to the dataset and displayed as *scores Dis1_1*), *State variable:0|Status, Value of state variable: 1, With diagonal reference line, Options, Test direction: Larger test include more positive test, Continue, Paste, Run selection*

Save *W_AnalysisSampleSyntax_9MDA.sps* and *W_AnalysisSampleOutput_9MDA.spo*. *Value of state variable* defines the value of the status variable representing 'positive' cases (defaults in our case study) to be put on the Y-axis of the ROC curve. In *Options* it is necessary to indicate, in accordance to the relative position of centroids, if higher values of scores indicate a higher or lower positivity (default risk in our case study). Figure 4.28 shows the results.

Horizontal and vertical straight segments of the ROC curve outline an accumulation of borrowers sharing the same score level (good and bad borrowers respectively). As a whole, the shape and the position of the curve outline an unsatisfactory discriminatory power of the model. In terms of AuROC, the model obtains 73.6% (corresponding to an AR of 47.2%). The confidence interval of AuROC, calculated at a confidence level of 95%, is in any case quite far away from the 50% that represents the naïve model. Therefore, as it is also indicated by the near zero p-value (Asymptotic Sig.), the model has a statistically significant discriminatory power. Remember that the best variable in terms of AuROC was DebtEquityTr (AuROC: 67%). So, the multivariate function has improved the AuROC above the level reached by the best individual variable of about six and an half percentage points. Incidentally, note that the cited variable has not been selected by the stepwise procedure that, on the contrary, chose ROEtr (whose

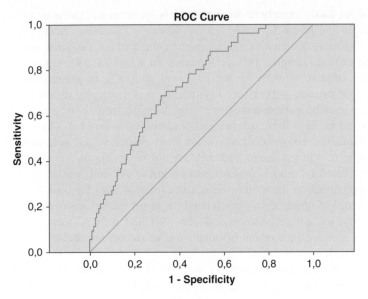

Area Under the Curve

Test Result Variable(s):Discriminant Scores from Function 1 for Analysis 1

			Asymptotic 95% Confidence Interval	
Area	Std.Error[a]	AsymptoticSig.[b]	Lower Bound	Upper Bound
0.736	0.032	0.000	0.674	0.798

[a] Under the nonparametric assumption
[b] Null hypothesis: true area = 0.5

Figure 4.28   ROC curve and AuROC.

AuROC is only 58%) at the first step. This happens because AuROC is a measure of the capability (of a variable or a function) to rank risk, whereas variable selection procedures also take into account the scale relation among a variable and the score.

We can now explore the role of prior probabilities. In the SPSS-PASW discriminant analysis procedure, in option *Classify*, we selected '*All groups equal*'. If we had chosen to define priors on the basis of the relative mix of good and bad borrowers in the sample (selecting *Compute from group sizes*), in the contingency matrix we would have observed a higher value of correct classification of good borrowers and a lower value of correct classification of defaulted borrowers. Scores do not change; it is the cut-off that changes in order to increase the probability to correctly classify most of the category of borrowers that is larger in size. The result of this choice is shown in Tables 4.41 and 4.42. It has increased the overall percentage of correct classifications, but also the type 1 error rate. Now, 94% of

Table 4.41   Prior probabilities for groups computed from group sizes.

Prior probabilities for groups			
01STATUS	Prior	Cases used in analysis	
		Unweighted	Weighted
performing	0.960	1216	1216
default	0.040	51	51
Total	1.000	1267	1267

Table 4.42   Classification results[b,c].

			01STATUS	Predicted group membership		Total
				performing	default	
Original	Count	performing	1221	0		1221
		default	48	3		51
	%	performing	100.0	0.0		100.0
		default	94.1	5.9		100.0
Cross-validated[a]	Count	performing	1221	0		1221
		default	48	3		51
	%	performing	100.0	0.0		100.0
		default	94.1	5.9		100.0

[a]Cross validation is done only for those cases in the analysis. In cross validation, each case is classified by the functions derived from all cases other than that case.
[b]96.2% of original grouped cases correctly classified.
[c]96.2% of cross-validated grouped cases correctly classified.

actual defaulted borrowers are predicted to be good, against the 70% we had when we used the balanced 50%–50% priors. Of course, as scores have not changed, the ROC curve built on the new score variable (*Dis1_2*) also does not change.
Save the new files as:

W_AnalysisSampleSyntax_10MDA

W_AnalysisSampleOutput_10MDA

In conclusion, the model we have estimated in this case study is concrete evidence of the need to carry out a series of improvements concerning variables (for instance, extending the application of the types of transformations we proposed), and concerning the multivariate function (for instance, changing the entrance and removal parameters, eliminating variables entered with a wrong sign or being highly correlated).

The evaluation of a discriminant analysis function encompasses many aspects. There are at least three areas of analysis:

1. *Checking the signs of coefficients, the significance of individual coefficients, and the coverage of relevant information categories:* coefficients' signs must be coherent with working hypothesis set for each indicator, and must be stable in the canonical discriminant function and in the structure matrix. Each variable included into the model should add a relevant and statistically significant contribution to the model. Variables should cover all typical perspectives of analysis on borrowers' creditworthiness.

2. *Discriminatory power of the scoring function:* performance indicators must be satisfactory.

3. *Stability of discriminatory power:* performance indicators should not deteriorate too much when moving from the analysis sample to validation samples.

4.6.3 Logistic regression

Once empirical data have been collected and datasets have been examined by the univariate and bivariate analyses, described above, model builders usually test different statistical procedures in order to find models that have the best results and consistency. Currently, LOGIT models are increasing their popularity thanks to their lower statistical requirements, as observed in Chapter 3. In order to explain how to build a LOGIT model using SPSS-PASW and, at the same time, to spot differences with linear discriminant analysis, we are going to estimate a model that uses exactly the same explanatory variables that were used in the last developed LDA model (Table 4.36). From the SPSS-PASW main menu (Figure 4.29):

> *Analyze, Regression, Binary Logistic, Dependent: BADGOOD, Covariates: (IEonLIABLITIES, EQUITYonPERMANENT_CAPITAL, ROAminusIEonTL, ROETr, INVENTORY_PERIOD3, EQUITYonPERMANENT_CAPITAL3, SALESBin2cl1), Categorical: SALESBin2cl1, Continue, Options: maximum iterations = 100, Continue, Method:Enter,OK .*

Note that, as we want to manually select which variables to include in the model, we use the Enter Method in the main window of logistic regression in SPSS-PASW. Therefore, no automatic variables selection will be carried out, and the model will include all variables listed as 'Covariates' (independent variables). In addition, note that it is necessary to specify which independent variables are categorical, by clicking on the option in the upper right corner of the window. In doing so, SPSS-PASW automatically recodes a multinomial categorical variable in many binary zero-one variables (dummy variables), one for each level of the categorical covariate; then, for each observation, SPSS-PASW assigns a value of one to the dummy variable if the observation is characterized by that level, and otherwise a value of zero (Table 4.43).

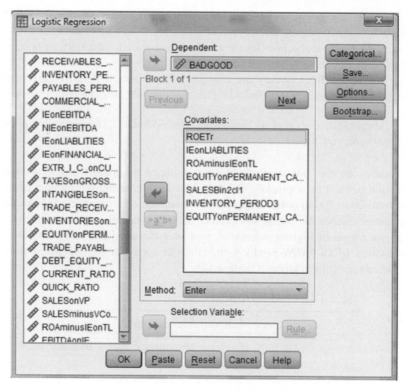

Figure 4.29   Logistic regression in SPSS-PASW.

Table 4.43   Example of a multinomial categorical variable recoding using dummy variables.

Original levels	Dummy A	Dummy B	Dummy C
A	1	0	0
B	0	1	0
C	0	0	1

However, it is necessary to exclude one of these *dummy* variables from the model in order to avoid *perfect multicollinearity* among dummy variables created from a multinomial categorical covariate. In fact, for each categorical variable with k mutually exclusive categories, it is enough to use $k-1$ *dummies* because the excluded category can always be identified by the zero in all other dummies. In a logistic model which includes one or more categorical covariates recoded as dummies, the circumstance in which all dummies are zero is known as '*baseline*'; as we shall see, the baseline is the reference point for interpreting coefficients estimated for independent categorical variables.

For each categorical variable, in the option 'Categorical' in the SPSS-PASW window for logistic regression, it is possible to select which level has to be excluded between the first and the last (levels are alpha-numerically ordered). Note that the method does not actually change until you click 'Change'.

In our model, there is only one categorical variable ('SALESBin2cl1') and it only has two categories; in order to include this variable as a categorical covariate:

Categorical, Categorical covariates: SALESBin2cl1, Reference category: First, Change, ok.

There is another box on the upper right side of the main window of logistic regression defined as 'Options'. This is useful for increasing the number of iterations from 20 to 100 in order to avoid that the convergence algorithm for parameter estimations reaches the maximum number of iterations and stops too early.

In the output of logistic regression, just after the usual summary table of processed cases, SPSS-PASW produces the tables of codes for the dependent variable and for categorical covariates (Table 4.44).

Table 4.44   SALES2Bin2cl1 coding.

Categorical variables codings		Frequency	Parameter (1)
SALESBin2cl1	0.00	883	0.000
	1.00	389	1.000

Subsequently, statistics relating to Step 0 in the procedure are reported; they refer to the 'null model', that is to say, the model including the intercept only (Table 4.45). Note that when using the value of the intercept (−3.176) in the equation of logistic regression, the result is the prior probability of default in the dataset (4%). It is usual, when estimating a model, to obtain negative values for the intercept because they must lead the dependent variable to typical samples' default rates, which are low.

Table 4.45   The null model.

Variables in the equation		B	S.E.	Wald	df	Sig.	Exp(B)
Step 0	Constant	−3.176	0.143	493.685	1	0.000	0.042

In Table 4.46 Chi-squared test results of the estimation procedure for Step 1 are reported. In this step, having requested to use the method Enter for variables,

Table 4.46   Chi-squared test on coefficients.

Omnibus tests of model coefficients				
		Chi-square	df	Sig.
Step 1	Step	47.539	7	0.000
	Block	47.539	7	0.000
	Model	47.539	7	0.000

all model variables (and the intercept) are considered. As usual, Sig. represents the test p-value; as it is much lower than the significance level which is commonly used (5%), the null hypothesis stating no effects of independent variables (globally considered) on the dependent variable can be rejected.

Table 4.47 indicates the estimate value for the log-likelihood function: this value is only useful when cross-comparing models built using different subsets of variables (*nested models*). To obtain summary measures of model performance, it is more useful to check the algorithm convergence within the maximum number of set iterations (see note in Table 4.47) and two statistics which indicate the value of two different 'pseudo R-Squared': Cox and Snell square and Nagelkerke square. In fact, logistic regression does not have an R-squared, but only indicators that are imitations of the R-square calculated for linear regression. There is a wide variety of pseudo R-squared. However, none of these statistics represents the proportion of explained variance, as R-square does in linear regression. Therefore, they must be interpreted cautiously, and should never be used alone when evaluating LOGIT models' performances.

Table 4.47   Model summary.

Model summary			
Step	−2 Log likelihood	Cox and Snell R-Square	Nagelkerke R-Square
1	380.473[a]	0.037	0.128

[a]Estimation terminated at iteration number 7 because parameter estimates changed by less than .001.

A very important SPSS-PASW output is shown in Table 4.48, which reports variables and coefficients of the estimated model. Analyze the table column by column.

Under the 'B' heading, estimated coefficients for the logistic regression function are reported. Each of these values indicates the sign of the relationship between a given independent variable and the dependent logit(π_i) function, that is $\log \frac{\pi_i}{1-\pi_i}$. In particular, a coefficient indicates the change of the logarithm of odds generated by a change of one unit in the explanatory variable, providing that other

Table 4.48   Variables in the equation.

	Variables in the equation					
	B	S.E.	Wald	df	Sig.	Exp(B)
Step 1[a] ROETr	−0.00005	0.000	0.156	1	0.693	1.000
IEonLIABLITIES	0.10245	0.054	3.541	1	0.060	1.108
INVENTORY_PERIOD3	0.00438	0.001	11.170	1	0.001	1.004
EQUITYonPERMANENT_ CAPITAL	0.00185	0.001	1.941	1	0.164	1.002
EQUITYonPERMANENT_ CAPITAL3	−0.00970	0.003	7.957	1	0.005	0.990
SALESBin2cl1(1)	−1.04512	0.419	6.232	1	0.013	0.352
ROAminusIEonTL	−0.00013	0.001	0.008	1	0.927	1.000
Constant	−3.54382	0.336	110.924	1	0.000	0.029

[a] Variable(s) entered on step 1: ROETr. IEonLIABLITIES. INVENTORY_PERIOD3. EQUI-TYonPERMANENT_CAPITAL. EQUITYonPERMANENT_CAPITAL3. SALESBin2cl1. ROAminusIEonTL.

independent variables remain constant. Considered as a whole, these parameters are the systematic component of the model:

$$-3,54382 - 0,00005 \cdot ROETr + 0,10245 \cdot IEonLIABILITIES + 0,00438$$
$$\cdot INVENTORY\_PERIOD3 + 0,00185 \cdot EQUITYonPERMANENT\_$$
$$CAPITAL - 0,0097 \cdot EQUITYonPERMANENT\_CAPITAL3$$
$$-1,04512 \cdot SALESBin2cl1(1) - 0,00013 \cdot ROA\min usIEonTL$$

As mentioned in Chapter 3, this equation corresponds to $\beta_0 + \sum_{j=1}^{p} \beta_j \cdot x_{ij}$, that is at the exponent of e in the following equation:

$$\pi_i = \frac{1}{1 + e^{-\left(\beta_0 + \sum_{j=1}^{p} \beta_j \cdot x_{ij}\right)}} \quad i = 1, \dots, n$$

Therefore, this last equation can be used to obtain the default probability π_i associated with each observation.

Under the 'S.E.' heading, the standard errors of estimated parameters are reported. Standard error can be used to verify if the parameter is statistically different from zero.

Under the 'Wald' and 'SIG.' headings, the Chi-square test and its p-value are reported; the null hypothesis that the coefficients are equal to zero is tested. Therefore, we hope to find p-values lower than the set level of significance (usually 5%) in order to reject the null hypothesis. In our model, the coefficients of ROETr, ROAminusIEonTL, ROETr, EQUITYonPERMANENT_CAPITAL and IEonLIABLITIES are not statistically significant at 5%.

Under 'DF', the degrees of freedom used for the test are indicated.

The column headed 'Exp (B)' reports estimated odds ratio of explanatory variables obtained by calculating the exponential of the parameter B. The borrower with a given level of a binary categorical variable represents the probability of default versus the probability of being non-defaulted. Therefore, as SALES-Bin2cl1 has an Exp(B) equal to 0.352, we can say that the odds of firms with sales lower than Euro 1,000 (SALESBin2cl1 = 1) is 0.352 times the odds of firms with sales larger than Euro 1,000 (SALESBin2cl1 = 0); as this odds ratio is lower than one, we can say that a negative relationship has been estimated between (the level 1 of) the categorical variable and the probability of default. For continuous variables, the odds ratio is the estimated relative risk due to the one-unit change in the continuous explicative variable: if an odds ratio is larger (lower) than one, the association is positive (negative) and the probability of default increases (decreases). This interpretation is, of course, in accordance with the indication obtained by considering the sign of B coefficients. There is also a relationship with the Wald statistic: if an odds ratio is close to one, the test is less significant, indicating a weaker relationship of the explanatory variable with the probability of default.

Our model (let's call it Logit Model 1) presents problematic results. Four out of the seven variables included in the logistic model (which were selected for the LDA model we are using as a benchmark) are not statistically significant. This signifies that obtaining a fair model by a given statistical procedure does not always represent a result that can be generalized. The estimated logit model has an AuROC similar to the benchmark discriminant model (Figure 4.30 and Table 4.49), but the presence of non-significant coefficients in the model requires further analyses.

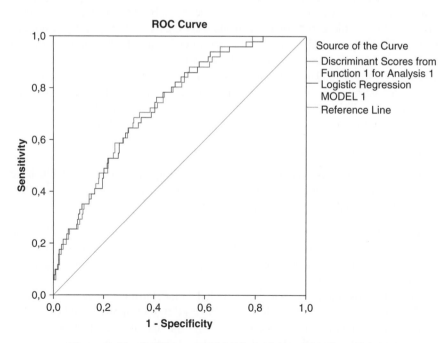

Figure 4.30   ROC curves of LDA model and Logit model 1.

Table 4.49   AuROC of LDA vs. logistic regression 1.

Area under the curve	
Test result variable(s)	Area
Discriminant Scores from Function 1	0.736
Logistic Regression MODEL 1	0.735

4.6.4   Refining models

Models can be refined from many perspectives. In this paragraph a few examples are given for logistic regression. In particular, four analyses are carried out, which relate to:

1. a priori selection of variables from the long list, followed by

2. a stepwise selection of variables,

3. expert-based refinement of logistic model,

4. transformations of variables.

4.6.4.1   New LOGIT models: selecting a short list of indicators and applying a stepwise procedure

In logistic regression, as in linear discriminant analysis, a stepwise procedure can be adopted to select variables. However, it is neither advisable nor feasible to apply this procedure to the entire long list of indicators, as we did when we developed the first discriminant analysis model. In fact, in estimating LOGIT models we can easily incur algorithm convergence problems when there are many variables that are heavily correlated. Therefore, it is necessary to select a short list of indicators. Univariate and bivariate analyses developed on indicators, and the summary statistics reported in Table 4.26, are useful for this purpose. Supposing we adopt the following criteria in selecting variables to enter into the short list:

- AuROC should be larger than 55%;

- original variables which have a corresponding transformed variable have been eliminated, thus privileging variables that, by the winsorization of the original variable, do not present outliers.

Table 4.50 shows the list of indicators that has been selected according to these criteria. From 73 variables in the long list, 25 variables have been entered into the short list. These variables have the best performance in terms of stand-alone AuROC and are winsorized.

Now, it is possible to estimate a new LOGIT model using a stepwise procedure to select the model's indicators. SPSS-PASW has different stepwise methods. They all share the same criterion for entering variables into the model, which is based on the calculation of a statistic (known as 'score statistic', not to be confused by the

Table 4.50   Variables in the short list and their AuROC.

Variable	AuROC
DebtEquityTr3	0.668
EBITDAonIE3	0.667
INVENTORY_PERIOD3	0.636
ASSETS_TURNOVER3	0.636
IEonEBITDA3	0.626
COMMERCIAL_WC_PERIOD3	0.624
TRADE_PAYABLESonTL3	0.621
ROAminusIEonTL3	0.619
QUICK_RATIO3	0.613
IEonLIABLITIES3	0.610
EQUITYonPERMANENT_CAPITAL3	0.606
NIEonEBITDA3	0.605
INVENTORIESonTA3	0.598
ROE3	0.596
EQUITYonLIABILITIES3	0.592
EBITDAonVP3	0.591
ROI3	0.587
ROA3	0.585
EBITDAonSALES3	0.577
ROS3	0.571
RECEIVABLES_PERIOD3	0.570
SALESBin2cl1	0.567
TRADE_RECEIVABLESonTA3	0.557
CURRENT_RATIO3	0.555
OVERSECT430_614	0.552

output of LDA models) for each variable and on its comparison with a maximum threshold. On the contrary, stepwise methods are diverse according to two elements: (i) removal criteria, which may be alternatively based on LR (Likelihood Ratio), Wald, Conditional; (ii) two iterative selection algorithms, which are:

- Backward. In this case, starting from a 'full model' that includes all available explanatory variables, these are removed one at a time, starting from the variable which has the largest value of the selected statistic criteria (LR, Wald, Conditional) beyond the threshold for variable removal. Once the removal process has terminated, variables which have a 'score statistic' below the chosen threshold are added to the model. The iterative process continues until the model satisfies all entry/removal requirements and stabilizes its structure.

- Forward. In this case, starting from the 'null model' (where only the intercept is included), variables that have a score statistic probability under the set threshold are entered one at a time (initiating from the most significant).

Then, the chosen removal test is performed and variables that have a statistic higher than the set threshold are removed; the iterative process continues until the model satisfies all entry/removal requirements and stabilizes its structure.

Choose the 'forward stepwise' based on LR removal criteria, with entry and removal threshold set as default by SPSS-PASW (respectively 5% and 10%). LR criterion is time consuming but tends to be more accurate than others. In any case, the three alternative removal criteria generally lead to the same results. From the SPSS-PASW main menu:

Analyze, Regression, Binary Logistic, Dependent: BADGOOD, Covariates: (All *variables in the short list), Categorical:* SALESBin2cl1, *Continue, Options: maximum iterations = 100, Continue, Method: Forward(LR),OK*.

The new model is reported in Table 4.51. It converges after five iterations and contains five variables, which are significant at 8%. Consequently, it is a more parsimonious model than the last LDA model we developed, which contained seven variables. The only common variable is SALESBin2cl1; economic and financial profiles considered by the two models are also quite different (Table 4.52). Coefficient signs are aligned with theoretical expectations. The AuROC of Logit Model 2 is slightly lower than LDA model's AurROC, 7.28% versus 7.36% (Figure 4.31 and Table 4.53).

Table 4.51   Logit model 2.

	Variables in the equation					
	B	S.E.	Wald	df	Sig.	Exp(B)
Step 5[a] EBITDAonIE3	−0.039	0.022	3.167	1	0.075	0.961
COMMERCIAL_WC_PERIOD3	0.002	0.001	4.347	1	0.037	1.002
TRADE_PAYABLESonTL3	−0.017	0.008	4.834	1	0.028	0.983
QUICK_RATIO3	−1.025	0.444	5.344	1	0.021	0.359
SALESBin2cl1(1)	−0.988	0.393	6.330	1	0.012	0.372
Constant	−1.600	0.397	16.269	1	0.000	0.202

[a]Variable(s) entered on Step 5: TRADE_PAYABLESonTL3.

To improve the AuROC performance of logistic regression we can, first of all, choose the backward stepwise procedure and, secondly, we can loosen entry and removal thresholds in order to increase the number of variables in the model. On one hand, the new model will be less parsimonious and variables will be less statistically significant; on the other hand, it can improve classification performance and increase the representativeness of different areas of analysis of a borrower's creditworthiness.

Table 4.52   Comparison of variables included into the last LDA model and Logit model 2.

LDA model variables	Logit Model 2 variables
ROETr	EBITDAonIE3
IEonLIABLITIES	
INVENTORY_PERIOD3	COMMERCIAL_WC_PERIOD3
EQUITYonPERMANENT_CAPITAL	QUICK_RATIO3
EQUITYonPERMANENT_CAPITAL3	TRADE_PAYABLESonTL3
SALESBin2cl1	SALESBin2cl1
ROAminusIEonTL	

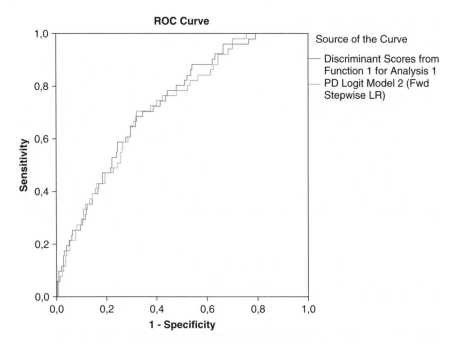

Figure 4.31   ROC curves of Logit model 2 and last LDA model.

Table 4.53   AuROC of Logit Model 2 and last LDA model.

Area under the curve	
Test result variable(s)	Area
Discriminant Scores from Function 1 for Analysis 1	0.736
PD Logit Model 2 (Fwd Stepwise LR)	0.728

Logit Model 3 is based on the backward stepwise algorithm with an entry threshold of 0.08 and a removal threshold of 0.012. Backward algorithm is by itself less selective because the removal threshold is dominating and is structurally higher than entry threshold. From the SPSS-PASW main menu:

Analyze, Regression, Binary Logistic, Dependent: BADGOOD, Covariates: (All *variables in the short list*), *Categorical:* SALESBin2cll, *Continue, Options: maximum iterations = 100, Probability of Stepwise: Entry (0,08) - Removal (0,12), Continue, Method: Backward(LR),OK*.

Results are reported in Table 4.54. The signs of variables coefficients are aligned with theoretical expectations.

Table 4.54   Logit Model 3.

		B	S.E.	Wald	df	Sig.	Exp(B)
				Variables in the equation			
Step 19	COMMERCIAL_WC_PERIOD3	0.003	0.001	6.068	1	0.014	1.003
	TRADE_PAYABLESonTL3	−0.015	0.008	3.345	1	0.067	0.985
	QUICK_RATIO3	−1.116	0.428	6.805	1	0.009	0.328
	IEonLIABLITIES3	0.103	0.060	2.962	1	0.085	1.108
	EQUITYonPERMANENT_CAPITAL3	−0.004	0.002	2.774	1	0.096	0.996
	EBITDAonSALES3	−0.018	0.011	2.699	1	0.100	0.982
	SALESBin2cll(1)	−0.890	0.394	5.098	1	0.024	0.411
	Constant	−1.967	0.540	13.253	1	0.000	0.140

This time, logistic regression needs 19 iterations to converge. The model includes seven variables, as did the last LDA model (Table 4.55). In this case, there are more common variables: two indicators are identical (EQUITYonPERMANENT_CAPITAL3 and SALESBin2cll), whilst one indicator (IEonLIABLITIES)

Table 4.55   Variables selection comparison: last LDA model vs. Logit Model 3.

Last LDA model's variables	Variables in Logit Model 3
ROETr	EBITDAonSALES3
IEonLIABLITIES	IEonLIABLITIES3
	TRADE_PAYABLESonTL3
INVENTORY_PERIOD3	COMMERCIAL_WC_PERIOD3
EQUITYonPERMANENT_CAPITAL	QUICK_RATIO3
EQUITYonPERMANENT_CAPITAL3	EQUITYonPERMANENT_CAPITAL3
SALESBin2cll	SALESBin2cll
ROAminusIEonTL	

Table 4.56   AuROC for last LDA model and Logit Model 3.

Area under the curve	
Test result variable(s)	Area
Discriminant Scores from Function 1 for Analysis 1	0.736
PD Logit Model 3 (Bwd Stepwise LR)	0.736

is selected in the original version in the last LDA model and in its winsorized transformation in Logit Model 3.

In terms of AuROC, the new model performance equalizes the benchmark LDA model (Table 4.56).

4.6.4.2   Experts-based refinement of the logistic model

Often, as stepwise selection of variables is a data-driven procedure, it is not sufficient to guarantee the model's optimization, both from a statistical point of view (for instance, due to multicolinearity phenomena) and from the economic meaningfulness of the model's point of view (for instance, due to excessive variables concentration in representing specific borrowers' profiles). Therefore, stepwise selection is useful to filter variables and to obtain evidence of the most important variables for a multivariate analysis. Nevertheless, a further refinement of the model is usually needed. It is typically based on experts' experience and knowledge of statistics and finance. To provide some examples of models' improvement, we are going to:

1. compare variables in Logit Model 3 with the short list of indicators in order to verify if important variables for financial analysis have been excluded;

2. verify the impact on AuROC when other variables are forced into the model or an original variable is substituted by its winsorized version included in the model;

3. check indicators representativeness and model parsimony.

By applying these judgement-based choices, a new model has been estimated using the 'Enter' procedure. From the SPSS-PASW main menu:

Analyze, Regression, Binary Logistic, Dependent: 01Status, Covariates: (EBITDAonIE3, INVENTORY_PERIOD3, IEonLIABLITIES, T2ROE, SALESBin2cl1, TRADE_PAYABLESonTL3, EQUITYon-PERMANENT_CAPITAL), *Categorical:* SALESBin2cl1, *Continue, Options: maximum iterations = 100, Continue, Method: Enter, OK*.

Logit Model 4 (Table 4.57) includes seven indicators, which are more significant in terms of the Wald test if compared, on average, to Logit Model 3. The model's

Table 4.57   Logit Model 4.

Variables in the Equation						
	B	S.E.	Wald	df	Sig.	Exp(B)
Step 1[a] EBITDAonIE3	−0.046	0.023	3.994	1	0.046	0.955
INVENTORY_PERIOD3	0.003	0.001	6.398	1	0.011	1.003
IEonLIABLITIES	0.062	0.034	3.426	1	0.064	1.064
T2ROE	4.216	2.068	4.154	1	0.042	67.743
SALESBin2cl1(1)	−1.153	0.421	7.507	1	0.006	0.316
TRADE_PAYABLESonTL3	−0.019	0.008	5.954	1	0.015	0.981
EQUITYonPERMANENT_CAPITAL	0.001	0.000	5.453	1	0.020	1.001
Constant	−4.825	1.158	17.371	1	0.000	0.008

[a]Variable(s) entered on Step 1: EBITDAonIE3. INVENTORY_PERIOD3. IEonLIABLI-TIES. T2ROE. SALESBin2cl1. TRADE_PAYABLESonTL3. EQUITYonPERMANENT_-CAPITAL.

equation is as follows:

$$\text{Logit}(\pi) = -4{,}82523 - 0{,}04627 \cdot \text{EBITDAonIE3} + 0{,}00341 \cdot$$

$$\text{INVENTORY\_PERIOD3} + 0{,}06242 \cdot \text{IEonLIABILITIES} + 4{,}21572 \cdot$$

$$\text{T2ROE} - 1{,}15339 \cdot \text{SALESBin2cl1}(1) - 0{,}0188 \cdot \text{TRADE\_PAYABLESonTL3}$$

$$+ 0{,}00075 \text{EQUITYonPERMANENT\_CAPITAL}$$

This new model has the highest AuROC (74.6%) in all models built in this case study (Table 4.58).

Table 4.58   AuROC values for different models under analysis.

Area under the curve	
Test result variable(s)	Area
Discriminant Scores from Function 1 for Analysis 1	0.736
PD Logit Model 3 (Bwd Stepwise LR)	0.736
PD Logit Model 4 (Judgemental-Based Selection)	0.746

However, as we have already mentioned a number of times, a single optimiza-tion criterion is not adequate. For instance, note from Figure 4.32 that the last model performs worse than other models when the assigned PDs are high (bottom left hand side of the figure); this is evidently more than balanced by better results for lower PDs (upper right hand side of the figure) as its AuROC is higher. As banks' lending policies typically differentiate risk appetite and pricing for different

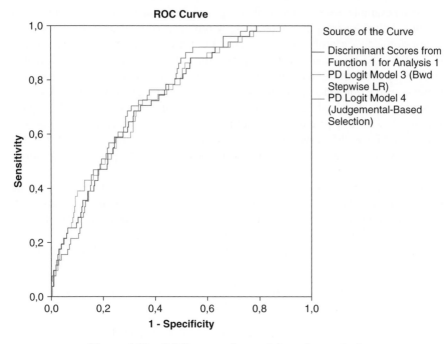

Figure 4.32   ROC curves for models under analysis.

levels of PDs, Logit Model 4 (that has weaker predictive capability for high PDs and stronger predictive capability for low PDs) should be judged according to banks' loan policies. In addition, note that a model can achieve the same AuROC in opposing circumstances (worse performance when assigned PDs are low and better results for higher PDs). Therefore, a model selection should be based on a set of valuations, many of which are judgment based.

One other thing to consider regarding Logit Model 4 is that coefficient signs are all aligned to expectations except in the case of T2ROE variable. This variable is a logistic transformation of ROE. As it is a profitability indicator, we would expect a negative sign for its coefficient, but we know that ROE is non-monotonic and this result could also be acceptable. However, there is no doubt that its inclusion in the regression should be considered as a weak aspect of the model.

4.6.4.3   Advanced variables transformations

Further refinements of multivariate statistical models may concern advanced transformation of individual variables. In Section 4.5.9, some transformations have already been presented in order to solve problems such as outlier values and non-normality or non-monotonicity of variables. Furthermore, advanced transformations of variables may be applied, with considerable gains in terms of models performance.

As an example, we examine nonlinear polynomial transformation. It is primarily useful to solve empirical non-monotonicity problems of indicators. Each indicator is transformed in a probability of default by using nonlinear regression. Results will tend to be monotonic with the probability of default itself, and thus highly performing in terms of predictive power. This transformation requires estimating a regression function that, thanks to its nonlinearity, may satisfactorily approximate PDs that are calculated for intervals of the original variable. Once the original variable has been segmented in a sufficient number of intervals, the empirical default rate for each interval is calculated and associated to the mean value of the interval (known as a node). To interpolate these nodes, an appropriate nonlinear function can be estimated.

A conceptually similar but more sophisticated approach is Local Polynomial Regression, which is able to model empirical default rates by a nonlinear regression, on a continuous basis. However, this function is not available in SPSS-PASW. Therefore, we can apply a different (but similar) approach.

All financial ratios included in the short list have been recoded as fractional rank-variables by:

> *Transform, Rank Cases, Variables* (All ratios in the short list), *Rank types: fractional rank, Continue, OK*.

This procedure produces a set of new variables (indicated by the prefix 'R') obtained by ranking each value of the variable; in other words, each value is transformed into a ratio which is calculated as: the rank of the value in the variable distribution divided by the overall number of values of the variable. Then, extreme ranked-values are substituted by 0.001 (for all values below this threshold) and by 0.999 (for all values over this threshold). Finally, for each of these variables, a new variable is produced (having 'PD_' as a prefix) by a nonlinear regression between the ranked variable used as explanatory variable and the dichotomous variable '01STATUS' used as a dependent variable. This nonlinear regression analysis estimates optimal parameters for the polynomial of degree n (with n no less than two). The larger the polynomial degree, the better the fit. Using ROE as an example, the syntax for a polynomial of degree four (which we use in order to transform all ratios) is:

MODEL PROGRAM A=0 B=0 C=0 D=0 E=0.
COMPUTE
PD_ROE=CDF.LOGISTIC(A*RROE**4+B*RROE**3+
C*RROE**2+D*RROE+E,0,1).
NLR BADGOOD
/PRED PD_ROE
/SAVE PRED
/CRITERIA SSCONVERGENCE 1E-8 PCON 1E-8.

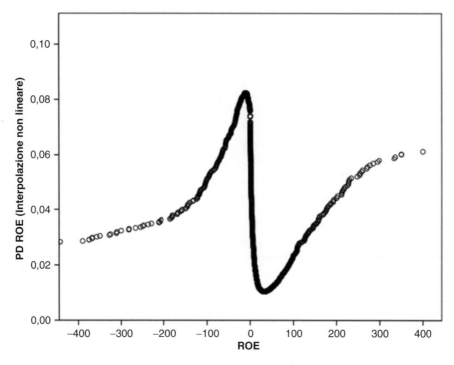

Figure 4.33   Nonlinear transformation of ROE.

Figure 4.33 shows the relationship between the original variable (ROE) and the transformed variable (RROE), which is an approximation of the probability of default calculated in local intervals. Remember that we have discussed a number of structural and economic determinants for ROE non-monotonicity. Therefore, this transformation should not be considered as a mathematical illusion, but rather as a solution for reflecting the true relation between ROE and the probability of default.

In fact, as shown by Figure 4.34 (left side), by segmenting the original ROE in five equal-percentiles intervals, its non-monotonic relationship with the probability of default is evident. Once the transformation has been carried out, the RROE appears monotonic (right side of Figure 4.34).

In terms of AuROC, transformed indicators show values that are, for the most part, higher than those of the original variables (Table 4.59).

At this point, we can estimate a fifth logistic regression model: we use the new transformed variables (without the nominal variables SALESBin2cl1, for the sake of simplicity) and adopt a forward stepwise procedure (with entry threshold

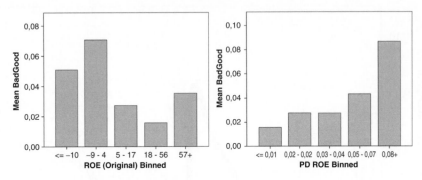

Figure 4.34   RROE vs. ROE: the relation with empirical default rates.

Table 4.59   AuROC values for original and transformed financial ratios.

Transformed variables	AuROC	Original variable's AuROC	AuROC difference
PD_ASSETS_TURNOVER	0.702	0.636	+6.6%
PD_TRADE_PAYABLESonTL	0.674	0.621	+5.3%
PD_COMMERCIAL_WC_PERIOD	0.670	0.624	+4.6%
PD_EBITDAonIE	0.666	0.666	+0.0%
PD_IEonEBITDA	0.664	0.626	+3.8%
PD_RECEIVABLES_PERIOD	0.662	0.570	+9.2%
PD_QUICK_RATIO	0.661	0.614	+4.7%
PD_DebtEquityTr	0.660	0.668	−0.8%
PD_ROE	0.658	0.594	+6.4%
PD_NIEonEBITDA	0.646	0.605	+4.1%
PD_INVENTORY_PERIOD	0.637	0.635	+0.2%
PD_ROAminusIEonTL	0.633	0.619	+1.4%
PD_IEonLIABLITIES	0.628	0.609	+1.9%
PD_TRADE_RECEIVABLESonTA	0.625	0.557	+6.8%
PD_EQUITYonPERMANENT_CAPITAL	0.607	0.606	+0.1%
PD_INVENTORIESonTA	0.606	0.598	+0.8%
PD_ROA	0.603	0.584	+1.9%
PD_EQUITYonLIABILITIES	0.598	0.593	+0.5%
PD_EBITDAonSALES	0.591	0.577	+1.4%
PD_EBITDAonVP	0.588	0.591	−0.3%
PD_CURRENT_RATIO	0.587	0.554	+3.3%
PD_ROI	0.584	0.587	−0.3%
PD_ROS	0.578	0.571	+0.7%

at 0.05 and a removal threshold at 0.10) in order to select the best combination of variables. The SPSS-PASW syntax of the new model is as following:

LOGISTIC REGRESSION VARIABLES BADGOOD
/METHOD = FSTEP(LR)
PD_EBITDAonIE PD_INVENTORY_PERIOD PD_IEonLIABLITIES PD_ROE
*PD_TRADE_PAYABLESonTL   PD_EQUITYonPERMANENT_CAPITAL   PD_*
DebtEquityTr
PD_ASSETS_TURNOVER PD_IEonEBITDA PD_COMMERCIAL_WC_PERIOD
*PD_ROAminusIEonTL   PD_QUICK_RATIO   PD_NIEonEBITDA   PD_*
INVENTORIESonTA
*PD_EQUITYonLIABILITIES   PD_EBITDAonVP   PD_ROI   PD_ROA   PD_*
EBITDAonSALES
PD_ROS PD_RECEIVABLES_PERIOD PD_TRADE_RECEIVABLESonTA
PD_CURRENT_RATIO
/save pred
/CLASSPLOT
/PRINT = GOODFIT CORR SUMMARY
/CRITERIA = PIN(0.05) POUT(0.10) ITERATE(100) CUT(0.5).

Table 4.60 reports the subsequent results. The algorithm converges after six iterations and the model includes six variables, which are all significant at 5%. Of course, all coefficients signs are positive, as variables indicate a probability of default by themselves.

Table 4.60   Logit Model 5.

Variables in the equation						
	B	S.E.	Wald	df	Sig.	Exp(B)
Step 6[a] PD_ROE	17.325	6.053	8.194	1	0.004	3.344E7
PD_TRADE_PAYABLESonTL	16.847	6.406	6.916	1	0.009	2.073E7
PD_ASSETS_TURNOVER	17.445	4.740	13.544	1	0.000	3.769E7
PD_QUICK_RATIO	26.070	8.094	10.373	1	0.001	2.099E11
PD_EBITDAonVP	25.896	12.652	4.189	1	0.041	1.764E11
PD_RECEIVABLES_PERIOD	17.965	5.551	10.473	1	0.001	6.340E7
Constant	−8.714	0.872	99.850	1	0.000	0.000

[a]Variable(s) entered on Step 6: PD_EBITDAonVP.

Figure 4.35 shows the ROC curve for the new model. The overall model performance in terms of AuROC is much higher than for previously examined models, reaching 84%, with an improvement of about 10 percentage points (Table 4.61).

It is worth noting that all built LOGIT Models largely focus on borrowers' profiles related to working capital. This can be considered as an over-specification

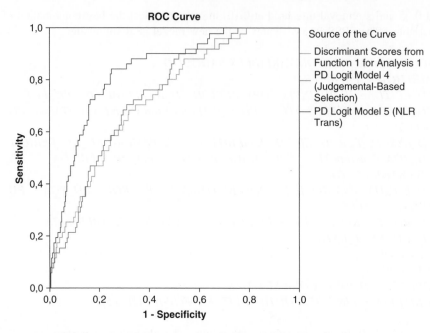

Figure 4.35   ROC curves for Logit model 5 and other models.

Table 4.61   AuROC values for Logit Model 5 and other models.

Area under the curve	
Test result variable(s)	Area
PD Logit Model 5 (NLR Trans)	0.840
PD Logit Model 4 (Judgemental-Based Selection)	0.746
Discriminant Scores from Function 1 for Analysis 1	0.736

of the models, possibly leading to poor performances in predicting defaults on longer time horizons. This is further evidence of the need to continue the model development beyond what is examined in this case study. Not to mention the need to implement all the steps of model calibration in order to reach consistent rating classes and probabilities of default estimates.

Save the output and the dataset with the new variables as:

W_CS_1_AnalysisSampleDataSet_11LR.spv

W_CS_1_AnalysisSampleDataSet_11LR.sav

4.7 From scores to ratings and from ratings to probabilities of default

Model quantification or calibration is the step devoted to associating rating classes and probabilities of default to the model's output. In the case of discriminant analysis, this step is essential because the model's output are scores that do not represent probabilities of default (even if scores can be transformed into probabilities; SPSS-PASW has a specific option in the section Save of the procedure Analyze, Classify, Discriminant that allows transforming them: all that is needed is to flag Probabilities of group membership). In general, the problem is that having rating classes permits a clearer straightforward quantification and validation of the model to be performed. In fact, if a single borrower, observed *ex post* (at the end of the observation period), can only either be good or bad, considering a rating class you can compare the estimated probability of default with the actual default rate of the class. In addition, there is a need to have stable judgments to communicate to customers.

In practice, we need to introduce many cut-offs in order to transform continuous scores into intervals, each representing a rating class. Basel II requires a minimum of seven classes (plus the class for defaults), for systems aiming at being validated for the IRB approaches stated in the first pillar of the regulation. Banks, for competitive reasons, tend to use more granular rating scales. In particular, in those segments in which it is needed to have a higher precision in setting default risk and risk prices (this is typical, for instance, of the larger corporate segments).

There are many ways to set cut-offs and related score intervals to associate to rating classes. The easiest of them, that we can assume as an exploratory approach, is that of selecting equal frequencies of borrowers for each score interval. In this case, finding cut-offs means to look at the chosen percentiles of the distribution (depending on the number of intervals required).

Consider the dataset W_CS_1_AnalysisSampleDataSet_8MDA and the last developed discriminant function. To obtain classes of scores from SPSS-PASW, from the main menu:

> *Transform, Visual Binning, Variable to bin: Discriminant Scores, Continue, Binned variable:* Rating, *Label variable:* Rating equalperc, *Make cutpoints, Equal percentiles, Number of cut-points:6, Apply, Label: from 1 to 7, Paste*

'Rating' is the name of the new variable. The chosen label for the new variable ('rating equal percentiles') indicates that ratings will be based on equal percentiles. The variable's levels range from one to seven, one being the best class (therefore, as scores in our function indicate default risk, one will be associated with the lowest scores).

To obtain the frequency of default for each class (to be interpreted as the probability of default) and tests of association between the categorical variables 01Status and Rating, use the SPSS-PASW main menu:

Analyze, Descriptive statistics, Crosstabls, row: 01Status*; Column:* Rating*; Statistics: Chisquared; Cells, Percentages: Columns, Continue, Paste, Run selection*

Table 4.62 shows that the seven rating classes obtained present monotonically increasing default rates (% within rating equal perc) and apparently well spaced out default rate values. From this point of view, the estimated discriminant function has a satisfactory capability to differentiate the average default rate of the overall sample (4%) in seven classes characterized by monotonically increasing default rates, ranging from 0% to 9.9%.

Table 4.62   Cross-tabulation between 01Status and Rating.

			\multicolumn{8}{c}{rating equal perc}							
			1	2	3	4	5	6	7	Total
01STATUS	performing	Count	182	180	178	175	173	170	163	1221
		% within rating equal perc	100	98.9	97.8	96.2	95.6	93.4	90.1	96.0
	default	Count	0	2	4	7	8	12	18	51
		% within rating equal perc	0.0	1.1	2.2	3.8	4.4	6.6	9.9	4.0
Total		Count	182	182	182	182	181	182	181	1272
		% within rating equal perc	100	100	100	100	100	100	100	100

01STATUS * rating equal perc Cross-tabulation

Table 4.63 shows results for the Chi-square test, which is largely used to check the statistical significance of the association between two categorical variables (01Status and Rating); the aim is to verify if rating classes appear different by chance. The null hypothesis is that classes have the same default rate; as the test is statistically significant (p-value is close to zero) we have to reject the null hypothesis and conclude that there is a statistical relevance of the association.

A further statistical analysis to carry out in order to check the quality of the calibration of the model is to observe the size of the overlaps of confidence intervals of default rates. To have confidence intervals using SPSS-PASW:

Analyze, Descriptive statistics, Explore, Dependent: 01Status*; Factor:* Rating*, Display: Statistics, Statistics: Descriptives 95% confidence level, Continue, Paste, Run selection,*

Table 4.63   Chi-square tests for the model.

Chi-square tests			
	Value	df	Asymp. Sig. (2-sided)
Pearson Chi-Square	32.977[a]	6	0.000
Likelihood Ratio	36.962	6	0.000
Linear-by-Linear Association	31.185	1	0.000
N of Valid Cases	1272		

[a]0 cells (0.0%) have expected count less than 5. The minimum expected count is 7.26.

In the output file, Double click on results, then Pivot, Pivoting trays, and finally invert the position on axes of Rating and Stat Type. What you obtain is reported in Table 4.64. It indicates that the weakness of the discriminatory power suggested by AuROC impacts on the width of confidence intervals of default rates: they are quite overlapped. Therefore, the calibration of the model also confirms conclusions reached when we were analyzing the capability of the model to rank borrowers: we have the same needs for further improvements on variables transformation and selection as well as on multivariate function estimation. Save the new files as:

W_AnalysisSampleSyntax_11MDAbin

W_AnalysisSampleOutput_11MDAbin

W_CS_1_AnalysisSampleDataSet_9MDAbin

Segmenting scores using the equal percentiles approach is simple, objective, and assures that any rating class is well populated. However, it has two limitations. The first is that classes cannot present monotonic default rates. In this case, sometimes modifying cut-offs permits the objective to be reached; at other times, it is necessary to merge contiguous classes. An important (and also regulatory) requirement is not to concentrate borrowers in a few classes and have other classes largely under-populated; in this case, some rating classes would appear to be more virtual then real. A second limitation is that probabilities of default obtained by the equal percentile approach are not comparable with agencies' long term default rates, which are usually adopted as a benchmark. In addition, as in the same bank there are different rating systems for different market segments, their rating class and their probability of default would not be immediately comparable. This is why banks define a 'master scale' that is a reference point for different internal systems and a link to agencies' ratings (Figure 4.36).

Each rating class of the master scale represents a target probability of default for classes of different internal rating systems. In this case, to calibrate the model means to identify cut-offs that minimize the distance between empirical default

Table 4.64 Confidence intervals and descriptive statistics of default rates of rating classes.

| | | Descriptives[a] | | | | | |
| | | | rating equal perc | | | | |
		2	3	4	5	6	7
01STATUS	Mean	Statistic					
		0.0110	0.0220	0.0385	0.0442	0.0659	0.0994
		Std. Error					
		0.00775	0.01090	0.01429	0.01532	0.01845	0.02231
	95% Confidence Interval for Mean	Lower Bound					
		−0.0043	0.0005	0.0103	0.0140	0.0295	0.0554
		Upper Bound					
		0.0263	0.0435	0.0667	0.0744	0.1023	0.1435
	5% Trimmed Mean	Statistic					
		0.0000	0.0000	0.0000	0.0000	0.0177	0.0549
	Median	Statistic					
		0.0000	0.0000	0.0000	0.0000	0.0000	0.0000
	Variance	Statistic					
		0.011	0.022	0.037	0.042	0.062	0.090
	Std. Deviation	Statistic					
		0.10454	0.14702	0.19284	0.20611	0.24885	0.30009
	Minimum	Statistic					
		0.00	0.00	0.00	0.00	0.00	0.00
	Maximum	Statistic					
		1.00	1.00	1.00	1.00	1.00	1.00
	Range	Statistic					
		1.00	1.00	1.00	1.00	1.00	1.00
	Interquartile Range	Statistic					
		0.00	0.00	0.00	0.00	0.00	0.00
	Skewness	Statistic					
		9.460	6.575	4.840	4.472	3.527	2.699
		Std. Error					
		0.180	0.180	0.180	0.181	0.180	0.181
	Kurtosis	Statistic					
		88.455	41.692	21.663	18.203	10.558	5.345
		Std. Error					
		0.358	0.358	0.358	0.359	0.358	0.359

[a]01STATUS is constant when rating equal perc = 1. It has been omitted.

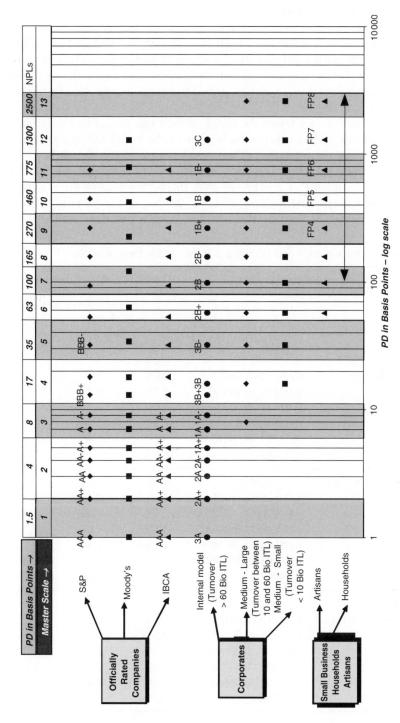

Figure 4.36   Master scale used by SanPaoloIMI in the earlier stages of internal ratings usage in the bank.

rates of internal rating classes and target probabilities of default defined on the master scale. Figure 4.36 suggests that the objective is largely achieved, notwithstanding the different granularity of internal rating systems developed for different market segments and the different rating assignment methodologies behind internal rating systems. Figure 4.36 also suggests that homologous rating classes of main international agencies are not always perfectly aligned among them in terms of probabilities of default.

The last issue is as follows: the sample default rate cannot be representative of the default rate of the market segment in which the model will be used for two reasons: (a) to aid the estimation of the model in case of low default portfolios, default proportion was artificially increased (b) historical datasets are based on a definition of default that is different from the current (and/or the regulatory) definition; this is typically the case of banks that are nowadays facing the new definition of default imposed by Basel II (that includes past due), while they did

Table 4.65   Default rates and confidence intervals for three models.

Equal percentiles rating classes		PD (%)	95% Confidence interval for mean	
			Lower bound (%)	Upper bound (%)
PD Logit Model 5	<= 0.00511	0.0%	–	–
(NLR Trans)	0.00512–0.00915	0.0%	–	–
(Binned)	0.00916–0.01440	2.7%	0.3%	5.1%
	0.01441–0.02239	0.0%	–	–
	0.02240–0.03740	1.7%	−0.2%	3.5%
	0.03741–0.07165	7.1%	3.4%	10.9%
	0.07166+	16.6%	11.1%	22.0%
Discriminant Scores	<=−0.64250	0.0%	–	–
(Binned)	−0.64249−−035335	1.1%	−0.4%	2.6%
	−0.35334−−012538	2.2%	0.0%	4.3%
	−0.12537–0.05643	3.8%	1.0%	6.7%
	0.05644–0.26200	4.4%	1.4%	7.4%
	0.26201–0.59696	6.6%	3.0%	10.2%
	0.59697+	9.9%	5.5%	14.3%
PD Logit Model 4	<=−0.00774	0.0%	–	–
(Judgemental-Based	0.00775–0.01660	1.6%	−0.2%	3.5%
Selection) (Binned)	0.01661–0.02559	0.5%	−0.5%	1.6%
	0.02560–0.03603	4.4%	1.4%	7.4%
	0.03604–0.04801	4.4%	1.4%	7.4%
	0.04802–0.06645	7.1%	3.4%	10.9%
	0.06646+	9.9%	5.5%	14.3%

not track past due phenomenon in the past and cannot reconstruct it today. In these cases, the calibration of probability of default of different rating classes by empirical default rates calculated on the sample would not lead to values representative of the true probabilities of default of the market segment. It is, therefore, necessary to rescale empirical probabilities to take into account the difference between the sample default rate and the market segment default rate which is aligned with the default definition that the bank intends to use.

A final analysis concerns the capacity of different models to provide adequate results in terms of PDs' differentiation for different rating classes. Using the same equal percentiles approach and seven classes for all models, we can compare the last discriminant model, Logit model 4 and Logit Model 5. By analyzing Table 4.65, important observations can be made. Models with higher AuROC values unnecessarily show better performance when rating classes are built and automatic binning procedures are used. In particular, the model with the highest AuROC has non-monotonic default rates per class and largely overlapped confidence intervals. This is further evidence of the fact that selecting the optimal model must:

- take account of many profiles at the same time,

- reach final results of models before having the possibility of making a selection.

5

Validating rating models

5.1 Validation profiles

Ratings systems validation scopes and steps are presented in this chapter, while in Chapter 6 a case study regarding the quantitative validation of a statistical based rating model is discussed. As a rating system 'comprises all of the methods, processes, controls, and data collection and IT systems that support the assessment of credit risk, the assignment of internal risk ratings, and the quantification of default and loss estimates' (Basel Committee, 2004, §394), it is clear that the validation scope is quite wide.

The validation of internal ratings is strictly required by the Basel Committee (2004, §530) for banks willing to opt for Internal Rating Based (IRB) approaches: 'banks must have a robust system in place to validate the accuracy and consistency of their internal models and modeling processes. A bank must demonstrate to its supervisor that the internal validation process enables it to assess the performance of its internal model and processes consistently and meaningfully'. However, the validation of an internal rating system is critical to the validation of the whole credit risk management system of a bank, both from a regulatory point of view and from a business management point of view.

It is crucial to the former perspective because capital adequacy depends on rating systems for banks adopting Internal Rating Based Approaches according to the Basel II regulation (the use of IRB approaches for the purposes of calculating capital requirements is subject to an explicit approval by national supervisory authorities and follows a 'supervisory validation' of rating systems). In addition, it is critical because Pillar II of Basel II is focused on the adequacy of risk management systems in order to safely and rationally manage the bank. It is also critical from the latter perspective because key decisions concerning individual

Developing, Validating and Using Internal Ratings: Methodologies and Case Studies   Giacomo De Laurentis,
Renato Maino and Luca Molteni   © 2010 John Wiley & Sons, Ltd

loans underwriting decisions as well as credit portfolio management decisions depend on rating systems.

Therefore, the difference in scope of 'regulatory validation' and of 'internal validation' is more apparent than real. In addition, consider that in order to be validated for regulatory purposes, a system has to be previously internally validated; on top of that, the technical contents of validation processes are very similar in both cases. These are reasons why we are going to use almost indifferent regulatory requirements as internal validation requirements.

On an ongoing basis, in the validation process, the bank has to verify the reliability of the results generated by the rating system and its continued consistency with regulatory requirements and operational needs. The validation instruments and methods are periodically reviewed also, and adjusted and updated to ensure that they remain appropriate in a context of continually evolving market variables and operating conditions. According to the 'proportionality principle', the scope and depth of quantitative and qualitative validation should be correlated with the type of credit portfolios examined, the overall complexity of the bank, and the stability of markets.

Rating systems must undergo a validation process consisting of a set of formal activities, instruments, and procedures for assessing the accuracy of the estimates of all material risk components and the predictive power of the overall performance system. The Basel II regulation states that: 'The institution shall have a regular cycle of model validation that includes monitoring of model performance and stability, review of model relationships, and testing of model outputs against outcomes.' (Basel Committee, 2004, §417). However, the same regulation underlines that the validation process lies not only on statistical comparisons of actual risk measures against the *ex ante* estimates, checking of parameter calibrations, benchmarking and stress tests, but also involves analyses of all the components of the internal rating system, including operational processes, controls, documentation, IT infrastructure, as well as an assessment of their overall consistency. Therefore, validation also requires the assessment of the model development process, with particular reference to the underlying logical structure and the methodological criteria supporting the risk parameter estimates.

Validation includes, too, the critical verification that the rating system is actually used (and how) in the various areas of bank operations. This is known as the 'use test', also required by Basel II and better specified in Basel Committee (2006). The results of the validation process need to be adequately documented and periodically submitted to the internal control functions and the governing bodies. The reports shall specifically address any problem areas.

Figure 5.1 gives an overview of the essential steps of rating systems validation.

In summary, the validation process has the key role of reviewing model building steps and application choices, detecting weaknesses and limitations, verifying the proper use of the system, and last, but not least, analyzing contingent solutions planned in case the robustness of the model falls or is lacking. Best practices have to be monitored to minimize misalignments of the whole process of internal credit risk management.

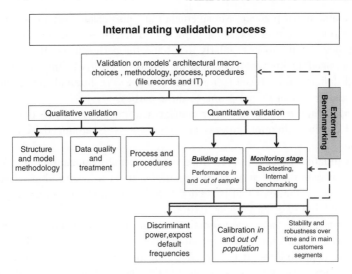

Figure 5.1   Fundamental steps in rating systems validation process.

5.2   Roles of internal validation units

The Basel II regulation is particularly innovative in terms of organizational requirements and internal controls. The rules lay down essential notions and criteria that banks must adopt in developing their rating systems. They also set down the organizational and quantitative requirements banks must comply with for recognition of their methods for capital adequacy purposes. The organizational requirements set rules which govern organization and controls, internal validation of rating systems, characteristics of rating systems (e.g., replicability, integrity, and consistency), their use in operations (use test), information systems and data flows. The quantitative requirements regard the structure of rating systems, the determination of risk parameters, stress tests, and the use of models developed by third-party vendors.

Specific requirements are set for the senior management and those who have roles in corporate governance and oversight.

'All material aspects of the rating and estimation processes must be approved by the bank's board of directors or a designated committee thereof and senior management. These parties must possess a general understanding of the bank's risk rating system and detailed comprehension of its associated management reports. Senior management must provide notice to the board of directors or a designated committee thereof of material changes or exceptions from established policies that will materially impact the operations of the bank's rating system' (Basel Committee, 2004, §438).

'Senior management also must have a good understanding of the rating system's design and operation, and must approve material differences between established procedure and actual practice. Management must also ensure, on an ongoing basis, that the rating system is operating properly. Management and staff in the credit control function must meet regularly to discuss the performance of the rating process, areas needing improvement, and the status of efforts to improve previously identified deficiencies' (Basel Committee, 2004, §439). Internal ratings must also be an essential part of the reporting to these parties.

In performing these tasks, senior management must consider recommendations produced by the validation process and review reports produced by the internal audit unit.

The validation process is performed by a specific organizational unit that may partially leverage on the support of operational units in performing its activities. In smaller banks, the least that is needed is the appointment of a manager devoted to coordinate and oversee these activities.

To perform these tasks, the validation unit has to be independent of other functions devoted to develop and to maintain model tools and to handle credit risk processes and procedures. It is advisable that the validation unit is also independent from those involved in assigning ratings and lending. Specifically, persons in charge of the function should not be subordinate to persons responsible for such activities.

Specific attention has to be paid to ensure the appropriate skills of human resources employed.

Where compliance with this requirement would prove to be excessively burdensome, the validation unit may be involved in the rating system design and development process, provided that appropriate organizational and procedural precautions are adopted and respected. In such a case, the internal audit function should verify that these activities are performed in an independent manner, fully achieving the intended objectives. The validation unit should also be independent from the internal audit function, which should review the validation process and findings.

In short, validation and control processes and organizational roles involved are depicted in Table 5.1.

Also, the internal audit function is deeply involved in validation processes, including the continued analysis of the compliance in the use of rating systems with internal and regulatory requirements. In particular, it is necessary to audit the independence of the validation unit and the quality of resources involved.

Validation is mostly performed on the basis of the documentation received by functions in charge of the model development and implementation in banks' credit processes. Therefore, the scope, transparency, and completeness of documentation are essential; these characteristics are important validation criteria. Banking groups with significant cross-border operations may have different organizational structures in different countries. Nevertheless, in all cases the parent company has to ensure that the organization of the validation and review functions within the group enable the unified management and control of models and rating systems.

Table 5.1   Processes and roles of validation and control of internal
rating systems.

	Models	Procedures	Tools	Management decision
Basic Controls	**Task**: model development and back testing **Owner**: credit risk models development unit	**Task**: credit risk procedures maintenance **Owner**: lending units / internal control units	**Task**: operations maintenance **Owner**: lending units / IT / internal audit	**Task**: lending policy applications **Owner**: central and decentralized units / internal control units
Second controls layer	**Task**: continuous test of models / processes / tools performance **Owner**: lending unit / internal audit		**Task**: lending policy suitability **Owner**: validation unit / internal audit	
Third controls layer	Risk Management/ CRO	Organisation / COO	Lending unit / CLO / COO	Lending unit – CLO / CRO
Accountability for supervisory purposes	Top management / Surveillance Board / Board of Directors			

CRO: Credit Risk Officer; CLO: Chief Lending Officer; COO: Chief Operating Officer;
IT: Information Technology Department

5.3   Qualitative and quantitative validation

There are two main areas of validation: qualitative and quantitative. Qualitative validation ensures the proper application of quantitative methods and the proper usage of ratings. Quantitative validation comprises all validation procedures of ratings in which statistical indicators are calculated and interpreted on the basis of an empirical dataset. In recent years, many books and articles have dealt with this topic, included among which are Engelmann and Rauhmeier (2006) and Christodoulakis and Satchell (2008).

Qualitative and quantitative validation complement each other. A rating procedure should only be applied in practice if it receives a positive assessment in the qualitative area. A positive assessment by the quantitative validation is not sufficient *per se*. Conversely, a negative quantitative assessment should not be considered decisive because statistical estimates are subject to random fluctuations and a certain degree of tolerance in the interpretation of results should be allowed. It is, therefore, necessary to place emphasis on qualitative validation.

5.3.1   Qualitative validation

5.3.1.1   Rating systems design

Rating systems design concerns the proper choice of the models architecture in relationship to the market segments in which the model is going to be used. It is necessary to ensure the transparency of the assumptions and/or evaluations which form the basis of the rating models design. The general suitability of a rating approach for specific rating segments has to be assessed. A number of other areas must be investigated:

- consistency of model development processes and methodologies,
- adequate calibration of model output to default probabilities,
- proper documentation of all model functions,
- analytical description of the rating process, with duties and responsibilities of key personnel,
- the robust procedures in place for validation and regular review.

In addition, there are important organizational profiles of rating systems' qualitative validation; they concern the link between the model, process, procedures, approval powers, and controls. Even the best model does not produce the expected added value to bank lending if it is misunderstood or if it is not adequately supported in daily applications. In this perspective, adequate education, clear procedures, proper guidelines, and support in tackling exceptions are fundamental. The assessment of the actual use of rating systems in credit approval processes is a key component of qualitative validation. In fact, the model must not only be a formal requirement for capital adequacy purposes or portfolio decisions; it must be fully integrated in the decision making process concerning single loans. If the bank credit culture does not accept the new model-based rating assignment processes, the risk of having two different processes (one being formal but inactive and the other informal but used in daily lending decisions) is very high. The validation has to detect these situations and suggest how to overcome them.

In the earlier stages of rating systems development in a bank, it commonly happens that credit risk functions spend a lot of time on model building, number crunching, statistical testing, and so on. Procedural aspects are underestimated in terms of the time, resources, and investments needed, as they are erroneously considered less problematic and easier to overcome. Since these early stages, the role of the validation unit in detecting the organizational readiness to accept and to correctly apply the new rating system is essential. The validation unit should have great visibility to top management and should lever on it in order to ask enough resources to properly take off the new process.

The essential requirements of rating systems that need to be checked in qualitative validation can be summarized in the following five main features:

- obtaining probabilities of default
- completeness

- objectivity

- acceptance

- consistency.

Obtaining probabilities of default   As mentioned in Chapter 2, ratings are the basis for almost all risk management applications once they have been quantified and probabilities of default have been obtained. In this perspective, different methods of rating assignment produce PDs in distinctive ways, as seen in Chapter 3. Statistical models are developed on the basis of an empirical dataset, which makes it possible to determine the PD for individual rating classes by calibrating results with the empirical data. Logistic regression enables the direct calculation of default probabilities, while for other methods (e.g., discriminant analysis) a specific adjustment is needed. Likewise, it is possible to validate the calibration of the rating model (*ex post*) using data gathered from the operational deployment of the model. Using this data, the default parameter can be constantly monitored and validated over time to maintain PDs aligned with real world outcomes.

Rating system completeness   Completeness is the next important feature of an internal rating system. In order to ensure the completeness of credit rating procedures, banks need to take all available information into account when assigning ratings to borrowers or transactions (Basel Committee, 2004, §417). The nature of the chosen rating assignment approach strongly impacts on this feature. Many default risk models use a small number of characteristics of the borrower to infer its creditworthiness. For this reason, it is important to verify the completeness of factors used to determine counterpart's creditworthiness, at least in model building stages and/or in the operational use (for instance, analyzing the scope of overrides proposed by credit analyst). In the estimation of statistical-based models, as a large number of borrowers' characteristics can be tested, the possibility to force variables to enter into the model in order to increase the completeness of the relevant risk factors should be verified. Usually, the computer-based processing of information enables expert systems and fuzzy logic systems to take a larger number of characteristics into consideration, meaning that such systems can be more comprehensive if properly modeled.

Rating system objectivity   A good rating system needs procedures that capture creditworthiness factors clearly and also minimize room for interpretation. Achieving high discriminatory power of ratings requires that they are assigned as objectively as possible, minimizing biases. In judgment-based approaches this can only be ensured by precise and plausible guidelines, common cultural backgrounds, appropriate training, ongoing benchmarking, and adequate organizational choices (team work, supervision, balancing individual analysts' specialization by sector, and analysts' teams' cross-sector mix). In statistical models, borrowers' characteristics are selected and weighed using an empirical dataset and objective methods; therefore, we can regard these models as the most 'objective' rating procedures.

When the model is fed by the same information, unavoidably the same results are obtained. This is also the case for expert systems and neural networks, where borrowers' creditworthiness is determined using defined algorithms and rules.

Rating system acceptance   Rating systems have also to be accepted by users, above all, internal users such as credit analysts, credit officers, and loan officers. Therefore, some requirements are necessary:

(a) The rating system should not produce classifications that are very often too far from those expected by bank analysts and officers;

(b) For small and medium enterprises, mechanical rating models often have higher discriminatory power than a poorly structured judgment-based approach developed by poorly experience and trained credit officers. However, they are less easily accepted because many actors do not have enough technical knowledge to understand them. Hence, an adequate education and level of disclosure on model frameworks for all actors involved in the lending process are indispensable.

Therefore, the validation process has to verify that rating models are well understood and shared by the users.

Different rating approaches have different degrees of acceptability. Generally speaking, as heuristic models are designed on the basis of experts' experience in lending, these models are more easily accepted; their credit assessments are considered warmer by end-users because they replicate their common culture. The acceptance of fuzzy logic systems may be lower as they require a greater degree of technical knowledge due to their fuzzy algorithms and changing variables' weights in different contexts. One severe disadvantage for the acceptance of artificial neural networks lies in their 'black box' nature. The increase in discriminatory power achieved by such methods depends on the network's ability to learn and on the parallel processing of information within the network. However, it is precisely this complexity which makes it difficult to comprehend results.

Rating system consistency   Consistency is the last but not least feature. Models have to be coherent and suitable for the borrowers to which they are applied and with the theoretical frameworks of users. When developing a statistical rating model, relationships between indicators may arise which contradict economic theory. Such contradictory indicators have to be excluded from further analyses; filtering out these problematic indicators serves to ensure consistency. Heuristic models do not contradict recognized scientific theories and methods, as these models are based on the experience and observations of credit experts. Pure statistical models depict business inter-relationships directly from empirical datasets and consistency should be checked.

The Basel II regulation states specific validation requirements in case statistical models and other mechanical methods are used to assign borrower or facility ratings or in estimation of PDs, LGDs, or EADs (Basel Committee, 2004, §417).

First of all, it is recognized that 'Although mechanical rating procedures may sometimes avoid some of the idiosyncratic errors made by rating systems in which human judgement plays a large role, mechanical use of limited information also is a source of rating errors. Credit scoring models and other mechanical procedures are permissible as the primary or partial basis of rating assignments, and may play a role in the estimation of loss characteristics. Sufficient human judgement and human oversight is necessary to ensure that all relevant and material information, including that which is outside the scope of the model, is also taken into consideration, and that the model is used appropriately'. This means that models must be part of a broader rating system, in which other methodologies add further information and expertise assuring completeness.

Other requirements of §417 are as follows: 'the burden is on the bank to satisfy its supervisor that a model or procedure has good predictive power and that regulatory capital requirements will not be distorted as a result of its use. The variables that are input to the model must form a reasonable set of predictors. The model must be accurate on average across the range of borrowers or facilities to which the bank is exposed and there must be no known material biases. The bank must have in place a process for vetting data inputs into a statistical default or loss prediction model which includes an assessment of the accuracy, completeness and appropriateness of the data specific to the assignment of an approved rating. The bank must demonstrate that the data used to build the model are representative of the population of the bank's actual borrowers or facilities. When combining model results with human judgement, judgements must take into account all relevant and material information not considered by the model. The bank must have written guidance describing how human judgement and model results are to be combined. The bank must have procedures for human review of model based rating assignments. Such procedures should focus on finding and limiting errors associated with known model weaknesses and must also include credible ongoing efforts to improve the model's performance ... The influence of individual factors on rating results should be comprehensible and in line with the current business research and practice. For example, if a multivariate statistical method is applied, factors in a statistical ratio analysis have to be plausible and comprehensible, according to the fundamentals of financial statement analysis and the economic theory of the firm.'

Therefore, in Paragraph 417 of the Basel II regulation, all five essential requirements (obtaining probabilities of default, completeness, objectivity, acceptance, consistency) for a satisfactory rating system have been detailed.

The same Basel II paragraph indicates two other important aspects of validation processes, that is to say, the continuity of validation processes and the completeness of documentation: 'The bank must have a regular cycle of model validation that includes monitoring of model performance and stability; review of model relationships; and testing of model outputs against outcomes ... In statistical models, special emphasis is to be placed on documenting the models statistical foundations, which have to be in line with the standards of quantitative validation.'

In examining all these features, the validation unit also has to take carefully into account external benchmarks, such as specialist literature and competitors application. The rating system is a decisional tool and can dramatically harm the bank's ability to compete if it is not aligned with those used by direct incumbents in the market.

5.3.1.2 Data quality

In statistical models, data quality is essential. Good data give outstanding results also using simple models, whereas the most advanced models cannot overcome poor data quality. Therefore, a comprehensive dataset is an essential prerequisite for quantitative validation. In this context, a number of qualitative aspects have to be considered:

- completeness of data,

- volume of available data,

- representativeness of samples used for model development and validation,

- consistency and integrity of data sources,

- adequacy of procedures used to ensure data cleansing and, in general, data quality.

The validation unit has a central role in confirming the dataset quality.

Particularly relevant are the reliability and completeness of defaulted observations because these are the actual limit to develop adequately large datasets for model development, rating quantification, and validation. The consistency of default definition used throughout data collection processes (that perhaps take place in different institution of a bank group, in different periods and countries) and its compliance with the Basel II definition of default (Basel Committee, 2004, §452) are both critical. Sample size is important as well as sample homogeneity: ideally, a sample has to be generated from a unique population using the same procedures, criteria, and methodology over the time. In other words, the sample must be generated by the same 'lending technology'. This is the set of information, rules, contracts, and policies applied to credit origination and monitoring; changing one or more of these components changes the credit portfolio generation and the borrowers' profile in the dataset (Berger and Udell, 2006) and can harm the consistency between the model development dataset and the population to which the model is operationally applied to.

A further profile of data quality is the time span to which data refers. Ideally, the dataset should be generated by considering an entire credit cycle; otherwise, estimates will be dependent on specific favorable or unfavorable cycle stages. As noted in Chapter 3, macroeconomic conditions are one of the most important determinants of default rates. If we miss a good representation of the credit cycle we miss something really relevant in describing default probability.

The combination of the last two mentioned conditions (lending technology stability and credit cycle coverage) proves to be very restrictive. We rarely observe procedures and processes that remain constant for five or more years of an entire credit cycle (the last started in 2002 and ended in 2008). Changes are more frequent because of the increasing technological opportunities to speed up processes and efficiency, discontinuities in the economic environment that lead to radically modifying credit policies, and new market segments becoming relevant; banks' mergers and acquisitions strongly impact on many aspects of the lending technology, too.

The validation process also has to pay attention to preliminary data treatment activities (such as finding and managing outliers, missing values, and poor data representativeness for some customers' segments).

Data quality is so relevant that the validation unit has to dedicate specific attention to these aspects. Figure 5.2 depicts the conceptual structure of data links from the model development dataset to the market the bank potentially confronts with.

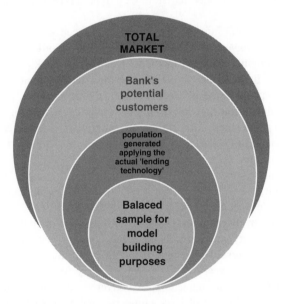

Figure 5.2   From samples to market population: data relationships.

Samples used in model building should have some desirable technical properties (low heteroscedasticity, no abnormal values, and so forth). Actual populations do not share these properties. The best way to extend a model's findings to populations is to apply a proper calibration and to perform *out-of-sample* analyses. These analyses are based on observations that are generated by the same lending technology but that were not included in the development sample. As a result, it is advisable to build various samples, one dedicated to support model building and others used for out-of-sample, out-of-time, and out-of universe validations of a model's performance.

The validation unit has an essential role in assessing two critical aspects: (i) stability of the lending technology behind data and (ii) proper model calibration in order to generalize results from sample to population. The two issues overlap, to some extent. If the observed in-sample default rate diverges from the total population, then calibration should reflect this divergence because the sample's central tendency would be different from the population's central tendency. This may simply be due to the fact that bank's lending technology is selecting borrowers better or worse than competing banks. This circumstance may also occur when lending technology changes: if the model is not re-calibrated, it continues to apply old criteria to new states of business. This is typically the case when mergers, acquisitions, demergers and so forth determine a change in the bank's lending technology.

The validation unit should be fully aware of the consequences of lending technology changes as well as of misalignments between borrowers' profiles in the original sample and population's profiles. If the rupture is significant, an extraordinary phase of model revision would be needed, at least in terms of model calibration.

Focus on calibration. Suppose that we use a balanced sample (50% performing, 50% defaulting borrowers) for model development in order to assure the best conditions for applying statistical methods: luckily, real banks' loan portfolios are much less risky. In other words, a normal long term annual default rate may be close to 2.5%; this value is far away from the 50% of the balanced sample. Moreover, defaults cluster together during the credit cycle with significant changes in default co-dependencies. The impact on calibration is significant; even small changes in model calibration have a big influence on a model's cut-off and on estimated default rates.

Figure 5.3 illustrates estimated PDs in a balanced sample, in a population where the default rate is 2.4%, and in a population whose default rate is 1% (for the calculation, see Section 3.2.4.3).

An inaccuracy in determining the long term average annual default rate modifies default probability measures. In fact, the lending process is relatively slow in producing evident results, also due to credit cycle movements. A credit cycle lasts years, not days or weeks. The central tendency (in statistics) is the average value to which population characteristics converge after many repetitions of the same process (this is the law of large numbers). Think about tossing a coin: after a few tosses, we cannot understand if the coin has been manipulated or not; we need a large number of trials in order to be sure that the coin is manipulated. The statistical repetitions in lending activities are relatively limited and it takes time to directly assess the effects of an incorrect parameter. Normally, a robust check on the validity of the central tendency is only possible after 18 or 36 months, depending on markets, types of facilities, and customers' segments.

In any case, the central tendency is a compromise between having long empirical series of observations and constant lending technology. Therefore, to set the central tendency is a very delicate issue that soon becomes a matter of discretion. The calibration turns into a managerial decision, which is partly based on empirical evidence and partly depends on strategies and policies (such as fixing the implicit 'risk appetite' of the organization). Optimistic estimates (default rate lower than actual) reduce the risk perception and determine aggressive competitive policies.

Calibration effects on model scores

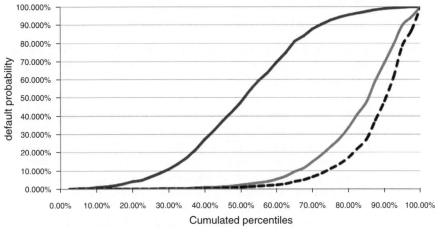

Figure 5.3   Calibration effects on model score estimated PDs using different long term average default rates.

If rating is also used for pricing purposes, then prices would not fully reflect the credit risk embedded in transactions (and loss provisions would be underestimated). On the contrary, if the estimated default rate is pessimistic, a conservative credit policy would be adopted, which would lead to missed business opportunities, to overestimated provisions, and to lower credit market shares.

In conclusion, the validation unit has an important role in verifying the central tendency over time through back testing and stress testing. It should carefully monitor market prices, signals from marketing people, results of big ticket transactions (syndicated loans, securities placing, securitisation, and so forth) and fully exploit any other opportunity to benchmark the bank (and models used) against direct competitors.

5.3.2   Quantitative validation

Quantitative validation covers four main areas:

1. Sample representativeness of the reference population at the time of the estimates and in subsequent periods.

2. Discriminatory power: the accuracy of ratings assignments in terms of the models' ability to rank obligors by risk levels, both in the overall sample and in its different breakdowns (for example, based on business sector, size and location).

3. Dynamic properties: the stability of rating systems and properties of migration matrices.

4. Calibration: the predictive power concerning probabilities of default.

We have already dealt with the issues of data quality extensively. Here we consider the perspective of samples size. Nowadays, the real constraint is usually given by the subsample size of defaulted firms, as some loan portfolios are characterized by very few defaults. As risks of these 'low-default portfolios' have to be assessed in any case, rating systems have to be developed and validated. A set of principles should be taken into consideration. Firstly, we cannot exclude exposures from the scope of application of the rating model simply because insufficient data are available to validate the risk parameter estimates on a statistical basis. In these cases, the validation unit has to contribute to set an adequate margin of conservatism in the assumption of risk parameters. Moreover, validation has to pay particular attention to analysis techniques adopted in this estimation process and to their limitations. Many statistical tests depend on the amount of available information. For instance, for the Chi-square test to give accurate results when dealing with contingency tables cross-tabulating a dichotomous variable, such as default/non-default with many rating classes, no more than 20% of cells should contain expected default frequencies less than five and no cells should have expected frequencies less than one. In many cases, minimum sample size requirements are not achieved, mainly due to the small number of defaults. This is particularly true when we are building models for market 'niches' or for specific industries (that are maybe important for their economic impact but that are composed by few competitors and counterparties). In these cases, we need to apply specific techniques to give more robustness to our estimates (Wehrspohn 2004, Basel Committee 2005b, Pluto and Tasche, 2004); among them, 'bootstrap procedures' have an important place. These procedures randomly generate many samples. Retaining the number of (the few) available defaults, many balanced samples can be iteratively generated by extracting an equal number of units from the non-defaulted group, without re-introduction. On each of these samples the rating model is completely re-assessed, extracting the entire set of statistical information (variables selected, means, standard deviations, likelihood tests, and so on). The set of models is then analyzed. If a clear convergence on a final stable result (i.e., same final variable selected, equivalent parameters, and so on) is found, we can infer that the model solution is stable and robust enough. If not, there would be a severe risk of instability and a more in-depth analysis would be needed. A way to overcome these problems is to find more homogenous subsets (applying cluster analysis, for instance). The model could be adapted to the specific features of these subsets, adopting different calibrations or integrating a specific successive qualitative analysis, maybe based on experts' judgments.

The term 'discriminatory power' refers to the fundamental ability of a rating model to differentiate between defaulting and performing borrowers over the forecasting horizon. Note that the forecasting horizon is usually set at 12 months for PD estimation (this also is a Basel II requirement) but the relevant time horizon for rating validation is the one set for rating assessment: in this last case, Basel II also requires a longer time horizon. Therefore, it is necessary to use longer forecasting horizons in order to validate discriminatory power. For example, the discriminatory power of a scoring model for installment loans is often calculated for the entire period of the credit transaction.

The discriminatory power of a model can only be reviewed ex post using data on defaulted and non-defaulted cases (back testing). Therefore, using a longer time horizon means using an 'observation period' that is more distant from 'time zero' and from the collection time of data which feeds model explanatory variables (Figure 4.1).

On the basis of the resulting sample, various analyses of the rating discriminatory power are possible. The list of methods in Basel Committee (2005a) is:

- statistical tests such as Fisher's r^2, Wilks' λ, Hosmer-Lemeshow;

- migration matrices;

- accuracy indexes such as Lorentz's concentration curves and Gini ratios (in different variants, for instance ROC and AuROC)

- classification tests (binomial test, type 1 and type 2 errors, χ^2 test, normality test and so forth).

The frequency distribution of good and bad cases is particularly important. In fact, error rates are the best way to offer a glimpse on model performances. The validation unit has to carefully verify the cut-off choice, its calibration, and its consequence in daily operations (as 'false good' cases create loss given default, and 'false bad' cases cause opportunity costs).

Ratings stability can be assessed by observing 'migration matrices'. They can be built once the rating system has been operational for at least two years. Desirable properties of annual migration matrices are:

- Transition rates to default should be in ascending order as rating classes worsen.

- High values should be on the diagonal and low values off-diagonal, which would signal that ratings are stable over time. This is also an indication of a through-the-cycle rating model, as opposed to point-in-time ratings, which are much more dynamic during the credit cycle, moving frequently from one class to another.

- Off-diagonal values should be in descending order when departing from the diagonal. That is to say, migration rates of plus or minus one class should be higher than migration rates of plus or minus two classes, and so forth. This means that rating movements are gradual whereas sudden leaps of many classes at one time are not that frequent.

These properties have to also hold for longer time horizons than one year, despite a natural reduction in on-diagonal values and an increase in off-diagonal values. This means that ratings change over time but without large leaps.

If analyses of firms' fundamentals are dominant in rating assignment, ratings change slowly over time because they are less sensitive to credit cycles and to transitory circumstances. Therefore, stability of the migration matrix is generally assumed as an indicator of an analytical process which is mainly centered on counterparty's fundamentals, and hence as an expression of a forward looking rating system.

This is a desirable technical property for many economic reasons, such as lower pro-cyclical effects (on banks, firms and, hence, on the economy as a whole) and longer 'far-sightedness' of credit allocation (Draghi, 2009). We return to these concepts in Chapter 7.

Calibration is a key topic in quantitative validation. It is also a critical issue because of the scarcity of statistical tools that are available. A document issued by the Basel Committee which is entirely dedicated to the validation of internal rating systems, clearly states that: 'compared with the evaluation of the discriminatory power, methods for validating calibration are at a much earlier stage ... Due to the limitations of using statistical tests to verify the accuracy of the calibration, benchmarking can be a valuable complementary tool for the validation of estimates for the risk components PD, LGD and EAD. Benchmarking involves the comparison of a bank's ratings or estimates to results from alternative sources. It is quite flexible in the sense that it gives banks and supervisors latitude to select appropriate benchmarks' (Basel Committee, 2005a, p.3).

Therefore, validating calibration means analyzing differences between forecasted PDs and realized default rates. The Basel Committee paper indicates a few tests to assess proper calibration: Binomial test, Chi-square test (or Hosmer–Lemeshow), Normal test, and Traffic lights approach. While the Binomial test is applied to one rating category at a time, the Chi-square test simultaneously checks several rating categories. The normal test is applied to a single rating class but is a multiperiod test of correctness of default probability forecasts; it is based on a normal approximation of the distribution of the time-averaged default rates (and on the assumptions that the mean default rate does not vary too much over time and that default events in different years are independent). The Traffic light approach is a multiperiod back testing tool for a single rating category introduced with the 1996 Market Risk Amendment as a supervisory evaluation tool of internal market risk models. Each of these tests bears important limitations. Therefore, we can conclude with the Basel Committee's words: 'at present no really powerful tests of adequate calibration are currently available' (Basel Committee, 2005a, p.34).

5.3.2.1 Back testing, benchmarking and stress testing

Back testing (accuracy of risk parameter estimates when compared with *ex post* empirical evidence), benchmarking (relative performance of systems and risk parameter estimates against benchmarks), and stress testing (adequacy of models when stress tests are applied) are three fundamental activities for validating rating systems.

When back testing, realized default rates must regularly be compared with estimated PDs for each rating grade. Where they do not fall within the expected range for that grade, the validation unit should analyze the reasons of deviations. Internal standards should be set for situations where deviations from expectations in realized PDs become significant enough to call the validity of estimates into question. These standards may take account of business cycles and similar systematic variability in default experiences. Where actual values continue to be higher than expected values, the bank should revise estimates upwards to reflect their default experience.

When benchmarking, the validation unit establishes procedures to specify acceptable deviations between internal estimates and benchmark data and identifies, at least in general terms, the actions to be taken when such deviations significantly exceed acceptable levels. Banks should also identify possible sources of unexpected volatility that could affect benchmarking results over time. This analysis should be conducted at least once a year. The adequacy and reliability of benchmarks is obviously critical. The comparisons of synthetic measures of rating performance must be carefully considered, as some very common indicators are sample dependent (such as Gini ratio and AuROC). It is much better to have benchmark datasets for testing different models on the same set of data.

Regarding a models' stress testing, the validation unit should assess the robustness and reliability of models' results when their independent variables are set to indicate extreme conditions.

Benchmarking, stress testing and, above all, back testing should be reported in an effective, easy to understand and transparent way to top managers. This would enhance the internal communication strategy of the validation unit: the clearer the communication, the more effective a top managers' contribution (to improve rating systems and to enhance rating validation activities) is.

As an example, suppose a bank has 15 000 internally rated customers; the internal rating system is based on 17 classes, without considering defaulted counterparties (Table 5.2).

Table 5.3 shows the loan portfolio by rating class at the beginning and at the end of the observation period. As indicated throughout this book, a number of performance measures and statistical tests can be calculated.

Effective and simple representation of this data is important to communicate to top managers and other bank personnel as well. Table 5.4 and Figure 5.4 illustrate a comparison between expected and actual default rates per rating classes. Deviations from means are highly frequent, mainly because of the effects of credit cycles. In periods of economic expansion, lower quality classes perform better than expected; the reverse would be true in periods of recessions. This is a well known phenomenon, well documented by rating agencies migration matrices observed in different periods.

When classes have few units, unexpected events hugely effect relative deviations but have a small economic impact (see class 3 for instance). The opposite is true for larger classes: even small deviations have a meaningful impact on portfolio performance. Therefore, these effects need to be carefully managed to avoid

Table 5.2   Internal rating classification.

	Probability of default (%)			Range (%)	
Rating class	Min	Mean	Max	Lower bound	Upper bound
1	0.01	0.03	0.04	−0.02	0.01
2	0.04	0.05	0.06	−0.01	0.01
3	0.06	0.07	0.08	−0.01	0.01
4	0.08	0.10	0.12	−0.02	0.02
5	0.12	0.15	0.19	−0.03	0.04
6	0.19	0.25	0.30	−0.05	0.05
7	0.30	0.40	0.50	−0.10	0.10
8	0.50	0.60	0.75	−0.10	0.15
9	0.75	0.90	1.15	−0.15	0.25
10	1.15	1.35	1.70	−0.20	0.35
11	1.70	2.00	2.50	−0.30	0.50
12	2.50	3.00	3.75	−0.50	0.75
13	3.75	4.50	5.50	−0.75	1.00
14	5.50	7.00	8.50	−1.50	1.50
15	8.50	10.00	13.00	−1.50	3.00
16	13.00	15.00	20.00	−2.00	5.00
17	20.00	25.00	50.00	−5.00	25.00

Table 5.3   Example of portfolio evolution in the observation period.

Rating classes #	Initial portfolio		Portfolio at observation period end				Frequency distribution by class (%)			
			Defaults		Non-defaulted		Default		Non-default	
	units	%		cumulated		cumulated		cumulated		cumulated
1	15	0.1	0	0	15	15	0.0	0.0	0.1	0.1
2	38	0.3	0	0	38	53	0.0	0.0	0.3	0.4
3	23	0.2	1	1	22	74	0.3	0.3	0.1	0.5
4	105	0.7	0	1	105	179	0.0	0.3	0.7	1.2
5	150	1.0	0	1	150	329	0.0	0.3	1.0	2.2
6	375	2.5	3	4	372	701	0.8	1.1	2.5	4.8
7	1170	7.8	4	8	1166	1.867	1.1	2.2	8.0	12.8
8	2138	14.3	6	14	2132	3.999	1.6	3.8	14.6	27.3
9	1725	11.5	5	19	1720	5.719	1.4	5.2	11.8	39.1
10	1650	11.0	15	34	1635	7.354	4.1	9.3	11.2	50.3
11	2100	14.0	32	66	2068	9.422	8.7	18.0	14.1	64.4
12	2250	15.0	55	121	2195	11.617	15.0	33.0	15.0	79.4
13	1200	8.0	56	177	1144	12.761	15.3	48.2	7.8	87.2
14	750	5.0	58	235	692	13.453	15.8	64.0	4.7	91.9
15	675	4.5	72	307	603	14.056	19.6	83.7	4.1	96.1
16	525	3.5	45	352	480	14.536	12.3	95.9	3.3	99.3
17	113	0.7	15	367	98	14.633	4.1	100.0	0.7	100.0
	15 000	100.0	367		14 633		100.0		100.0	

Table 5.4 Example of actual values against expected values in a portfolio during a favorable credit cycle.

Rating classes #	Central PD (%)	Defaults expected	Actual defaults	Default rate (%) Actual	Δ Actual versus expected	Survival rate (%)
1	0.03	0	0	0.0	0.0	100.0
2	0.05	0	0	0.0	0.0	100.0
3	0.07	0	1	4.4	4.4	95.6
4	0.10	0	0	0.0	−0.1	100.0
5	0.15	0	0	0.0	−0.2	100.0
6	0.25	1	3	0.8	0.6	99.2
7	0.40	5	4	0.3	−0.1	99.7
8	0.60	13	6	0.3	−0.3	99.7
9	0.90	16	5	0.3	−0.6	99.7
10	1.35	22	15	0.9	−0.4	99.1
11	2.00	42	32	1.5	−0.5	98.5
12	3.00	68	55	2.4	−0.6	97.6
13	4.50	54	56	4.7	0.2	95.3
14	7.00	53	58	7.7	0.7	92.3
15	10.00	68	72	10.7	0.7	89.3
16	15.00	79	45	8.6	−6.4	91.4
17	25.00	28	15	13.3	−11.7	86.7
		447	367	2.4		97.6

miscommunication (from this perspective, indicators like ROC curve are particularly suitable).

Linking crude data of rating classifications to bank's lending policy is useful for managers and for effective communication. Figure 5.5 offers a way to illustrate this analysis. On the graph the frequency distributions of actual defaulted and non-defaulted counterparts are shown. Of course, the two groups have different distributions and there is a large overlapping area. Rating classes are often the main drivers for bank lending policies. Different commercial policies are put into practice in respect of counterparty's credit risk, favoring aggressive marketing for safer clients and conservative lending behaviors for riskier ones. Suppose that aggressive marketing is pursued for better classes up to class 6, while a conservative approach is recommended from class 14 onwards. This policy neither protects against defaults in classes that benefit from aggressive marketing, nor avoids restricting lending to solvent counterparties. In our example, the target for aggressive marketing is around 700 clients (the first 5% of the portfolio) but three defaults were experimented (the first 1.1% of total defaults); see the gray area on the left in Figure 5.5. At the same time, if we withdraw credit to the worst three classes, 130 defaults could be avoided but business with 1200 clients would be lost (gray area on the right in Figure 5.5).

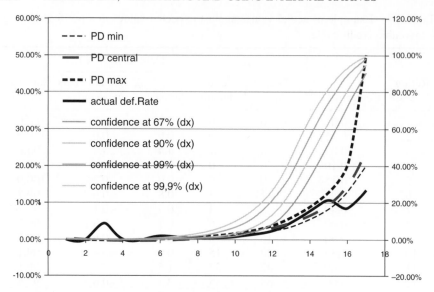

Figure 5.4   Default rates per rating class and statistical confidence intervals.

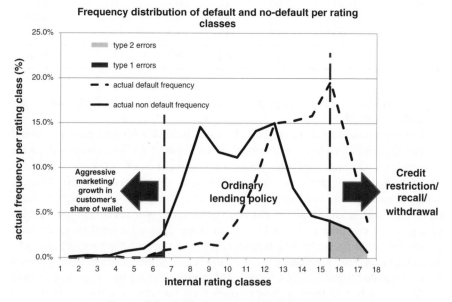

Figure 5.5   Default rates and lending policy.

The importance of a models' discriminatory power and adequate calibration becomes evidently clear. The usefulness of having clues on these performance measures of rating systems becomes apparent. Also, the value of a prompt detection of fading discriminatory power and calibration becomes evident.

6

Case study: Validating PanAlp Bank's statistical-based rating system for financial institutions

6.1 Case study objectives and context

Risk management units and their consultants are looking into building rating models with a good degree of robustness, strong performance and Basel II compliancy. Validation units must review the adequacy of datasets, methodologies, and performance. Internal auditing units evaluate the accuracy of ratings development and validation processes, in addition to their daily usage in operations in banks. Eventually, supervisory authorities must verify the adequacy of banks' credit risk management systems in light of the requirements set by the Second Pillar and, for banks applying for (foundation or advanced) internal ratings-based approaches, of the requirements set by the First Pillar.

This case study presents a 'development report' describing the estimate of a statistical-based scoring system, choices made, and the results obtained. The reader should take on the perspective of a validation unit whose task is to write a 'validation report' outlining acceptable choices, problematic issues, errors and omissions of the entire process of model building as detailed in the development report.

In particular, the following case study objectives are discussed:

(a) The coherence of steps sequence and the consistency of the overall process.

(b) The adequacy of data analysis, methodologies, and development choices.

Developing, Validating and Using Internal Ratings: Methodologies and Case Studies Giacomo De Laurentis, Renato Maino and Luca Molteni © 2010 John Wiley & Sons, Ltd

(c) The integrity and completeness of data used.

(d) The level of detail of the report.

The rating model has been built using the shadow ratings approach (that is, trying to replicate ratings assigned by rating agencies), since there is an insufficient number of default cases among the financial institutions financed by a bank. This peculiarity does not change the development and validation processes in a material way. Therefore, the case study can be generalized to almost any other segment of borrowers and to more typical modeling approaches. On the contrary, to consider a dichotomous variable different from the typical default/non-default case and a market segment different from the common corporate/SME/retail segments would be useful in order to increase the abstraction and generalization of the following discussion.

PanAlp Bank decided to adopt statistical-based rating systems in order to assign 'base ratings', to be subsequently completed by an override process (that is adding judgmental analyses by credit analysts). Until now, PanAlp Bank has focused on main segments of the loan portfolio, but recently their risk management unit has expanded with the addition of recently graduated young personnel and, subsequently, new models for other market segments have been developed. Moreover, the validation unit has grown in size, and a completely new team has been assembled. The first task assigned to this new team is the validation of a new model for assigning ratings to financial institutions.

Any assessor should, first of all, carefully read the 'development report'. Only the statistical model is under scrutiny, from a methodological and performance point of view. Organizational aspects are beyond the scope of this analysis. Methodology assessment also implicates compliance with the regulatory requirements of the Advanced Internal Rating Based Approach of Basel II, as PanAlp Bank is looking forward to applying such an approach to capital adequacy. The validation unit will identify elements of the development process to improve errors and omissions in the reporting. In addition, the unit intends to propose a general framework for model development and validation reporting to be used as a guideline and a learning tool for the future.

6.2   The 'Development report' for the validation unit

6.2.1   Shadow rating approach for financial institutions

To build a statistical-based scoring system for financial institutions, accounting data has been extracted from the Bankscope database. Bankscope is a global database containing data on about 25 000 listed and non-listed banks, located in many countries all over the world. For each bank, the following data are available:

- Balance sheets,

- Financial ratios,

- Rating issued by four international rating agencies,

- Ownership structure,

- Securities issued and their prices.

Given the limited number of defaulted banks, the shadow ratings approach has been adopted (that is to say, building a scoring model that replicates ratings issued by external credit assessment institutions as much as possible). Once the model has been estimated and its performance has been verified, it can be extended to non-rated banks. The dataset contains 56 banks in a time span of three years (2003, 2004, 2005), for a total of 168 observations. Of these observations, 33 (11 each year) are non-investment grade banks, and 135 observations (45 per year) are investment grade. This distinction is used in order to have a 0/1 dichotomous variable. Investment grade banks are indicated as 1; bad and good are two terms that are used to indicate non-investment and investment banks respectively; none of the banks considered has migrated from one to the other of these two classes.

For each bank, 30 of the 36 financial indicators calculated by Bankscope have been included in the dataset.

Once the dataset had been obtained, the following analyses were performed:

1. Missing values,

2. Economic meaning of indicators,

3. Monotonic relationship with risk,

4. Sign of the relationship with the probability of default (PD),

5. Normality of distributions,

6. Outliers,

7. Homogeneity of variance,

8. Discriminatory power of individual ratios,

9. Ratios correlations.

The result of these analyses is a short list of indicators considered potentially useful to enter a discriminant function. Using the stepwise method, a second level of ratios selection has been undertaken. Finally, the model has been estimated. Then, the model's performances have been analyzed using contingency tables and ROC curves.

6.2.2 Missing value analysis

Notwithstanding the care in extracting data, the dataset contains some missing values. Missing value analysis was performed at the outset in order to avoid repetition of other analyses after managing missing values. After having analyzed missing values, we decided to substitute them with the average value of the indicator for each of the two groups of investment/non-investment grades. From this point forward, all subsequent analyses were performed on the new dataset 'after missing value substitution': the resulting financial ratios are indicated in SPSS-PASW outputs as SMEAN(. . .).

6.2.3   Interpreting financial ratios for financial institutions and setting working hypotheses

Each considered ratio has an economic meaning clearly deducible by its structure. The working hypothesis is the expected sign of the relationship with the probability of default.

1. Loan Loss Reserves / Gross Loans: the ratio between the stock of cumulated provisions on loans and outstanding loans. Setting the working hypothesis for this ratio is not clear cut. Usually, it is assumed that a higher percentage of loan loss reserves make a bank safer. However, the possibility that higher loan loss reserves are simply reflecting riskier assets must be considered. Accepting the prevailing view, a negative sign is expected.

2. Loan Loss Provisions / Net Interest Revenue: the ratio between the annual provisions on loans and revenues from interest-earning assets, fees on loans and derivative contracts. A negative sign for the relationship with the probability of default of the bank is generally expected, reflecting a more conservative approach to show up profit.

3. Loan Loss Reserves / Impaired Loans: the ratio between the stock of cumulated provisions on loans and substandard loans. A negative sign for the relationship with the probability of default of the bank is expected, since a higher share of impaired loans is covered by reserves.

4. Impaired Loans / Gross Loans: the ratio of non-performing loans against outstanding loans. A positive sign is expected, due to the higher risk in loan portfolio.

5. Total Capital Ratio: regulatory capital divided by total risk-weighted assets. A negative sign for the relationship with the probability of default of the bank is expected because of the lower leverage.

6. Equity / Total Assets: accounting capital divided by total assets. A negative sign for the relationship with the probability of default of the bank is expected because of the lower leverage.

7. Equity / Net Loans: accounting capital divided by outstanding loans net of cumulated provisions. A negative sign for the relationship with the probability of default of the bank is expected because of the higher capital available per Euro of loan.

8. Equity / Deposits and Short Term Funding: accounting capital divided by deposits and other short term debt. A negative sign for the relationship

with the probability of default of the bank is expected because of the lower leverage.

9. Equity / Liabilities: accounting capital divided by liabilities. A negative sign for the relationship with the probability of default of the bank is expected because of the lower leverage.

10. Capital Funds / Total Assets: the ratio between the overall capital funds and total assets. A negative sign for the relationship with the probability of default of the bank is expected because of the lower leverage.

11. Capital Funds / Net Loans: the ratio between the overall capital funds and outstanding loans net of cumulated provisions. A negative sign for the relationship with the probability of default of the bank is expected because of the lower leverage.

12. Capital Funds / Deposits & Short-Term Funding: the ratio between the overall capital funds and deposits and other short term debt. A negative sign of the relation with the probability of default of the bank is expected because of the lower leverage.

13. Capital Funds / Liabilities: the ratio between the overall capital funds and liabilities. A negative sign of the relation with the probability of default of the bank is expected because of the lower leverage.

14. Subordinated Debt / Capital Funds: the ratio between subordinated (junior) debt and overall capital funds. A positive sign for the relationship with the probability of default of the bank is expected because of the higher share of capital funds made by subordinated debt rather than by more equity-like forms of capital.

15. Net Interest Margin: interest revenues minus interest costs in percentage of average earning assets (average of current and previous year assets). A negative sign for the relationship with the probability of default is expected because of higher profitability of the bank.

16. Net Interest Revenue / Average Assets: revenues from interest-earning assets, fees on loans and derivative contracts divided by average assets (average of current and previous year assets). A negative sign for the relationship with the probability of default is expected because of higher profitability of the bank.

17. Other Operating Income / Average Assets: income other than revenues from interest-earning assets, fees on loans and derivative contracts divided by the average assets. A negative sign for the relationship with the probability of default is expected because of higher profitability of the bank.

18. Non-Interest Expense / Average Assets: the ratio between 'expenses other than interest expenses' and average assets. A positive sign for the relationship with the probability of default is expected because of the burden of higher costs that the bank withstands.

19. Pre-Tax Operating Income / Average Assets: gross operating return on (average) assets. A negative sign for the relationship with the probability of default is expected because of higher profitability of the bank.

20. Non-Operating Items and Taxes / Average Assets: economic impact of non-operating items and taxes on average assets. Setting the working hypothesis for this ratio is not straightforward because it is difficult to link some components of the numerator (such as extraordinary items and taxes) with the probability of default. In general, considering the ratio as an indicator of profitability, a negative sign for the relationship with the probability of default can be set.

21. Return on Average Assets (ROAA): net income divided by average assets. A negative sign for the relationship with the probability of default is expected because of higher profitability.

22. Return on Average Equity (ROAE): net income divided by average equity (average of current and previous year equity). A negative sign for the relationship with the probability of default is expected because of higher profitability.

23. Non-Operating Items / Net Income: share of net income due to non-operating items. As in non-operating items there could be many non-recurrent revenues, the higher the share of net income made up by non-operating items, the higher the probability of default.

24. Cost to Income Ratio: operating costs divided by intermediation income. A positive sign for the relationship with the probability of default is expected because of the burden of higher costs that the bank withstands.

25. Recurring Earning Power: operating revenues before provisions divided by total assets. A negative sign for the relationship with the probability of default is expected because of higher profitability.

26. Net Loans / Total Assets: outstanding loans net of cumulated provisions divided by total assets. Loans are the riskier asset class in a bank portfolio, so the higher the share of loans is, the higher the overall bank risk is.

27. Net Loans / Customer and Short Term Funding: outstanding loans net of cumulated provisions divided by customers' funds and other short term debts. A positive sign for the relationship with the probability of default is expected, for the same reason given for the previously mentioned ratio.

28. Net Loans / Total Deposits and Borrowings: outstanding loans net of cumulated provisions divided by deposits and other debts. A positive sign for the relationship with the probability of default is expected, for the same reason given for the ratio no.26.

29. Liquid Assets / Customer and Short Term Funding: liquid assets in percentage of customers' funds and other short term debts. A negative sign for

the relationship with the probability of default is expected due to higher liquidity of the bank.

30. Liquid Assets / Total Deposits and Borrowings: liquid assets in percentage of deposits and other debts. A negative sign for the relationship with the probability of default is expected due to higher liquidity of the bank.

6.2.4   Monotonicity

A ratio is monotonically related to the probability of default if it shifts in the same direction with the PD for all its possible values. That is to say, an increase in the ratio value always represents an increase or a decrease of the probability of default. Many statistical techniques used to estimate statistical-based scoring systems require this characteristic for explanatory variables. All ratios in the dataset are monotonic, except 'Non-operating items / Net income'.

6.2.5   Analysis of means

A first check of working hypotheses set before can be developed by comparing the average values of each financial ratio for the two subsamples of good and bad borrowers. If the relationship of a given ratio with the probability of default is expected to be positive (negative), non-investment banks' mean should be higher (lower) than that of investment banks.

The vast majority of indicators behave as expected (Table 6.1).

6.2.6   Assessing normality of distributions: histograms and normal Q–Q plots

Some multivariate statistical techniques, such as Linear Discriminant Analysis, require explanatory variables be distributed as a normal distribution. To validate whether this requirement is satisfied, there are at least two alternatives: (i) graphical observation, to be subjectively interpreted and (ii) statistical tests.

In this section, graphical tools are introduced. In specific, there are two types of graphics:

1. Histograms, which provide a graphical representation of the frequency distribution of values of the examined financial ratio.

2. Normal Q-Q Plots, which compare the frequency distribution of the examined financial ratio with a normal distribution, which is indicated by a straight line.

These analyses have a certain degree of subjectivity, as it is not simple to define a clear divide between normality and non-normality. For instance, Cost to Income Ratio in Figure 6.1 has been considered to be normally distributed in both the bad and good subsamples. Equity / Total Assets in Figure 6.2 has been considered non-normal in both subsamples.

Table 6.1   Financial ratios means for non-investment grade and investment grade banks.

#	Indicator	Mean for non-investment grade banks (Status 0)	Mean for investment grade banks (Status 1)
1	Loan Loss Reserve/Gross Loans	6.6175	2.5416
2	Loan Loss Provisions/Net Interest Revenues	20.1266	6.6665
3	Loan Loss Reserve/Impaired Loans	106.6789	183.2355
4	Impaired Loans/Gross Loans	7.1836	2.4883
5	Total Capital Ratio	17.2371	17.7501
6	Equity/Total Assets	10.4725	13.1503
7	Equity/Net Loans	32.4994	18.3102
8	Equity/Dep & ST Funding	15.1203	57.4297
9	Equity/Liabilities	12.1601	19.4195
10	Cap Funds/Total Assets	10.1517	9.3306
11	Cap Funds/Net Loans (s)	32.7668	17.1826
12	Cap Funds/Dep & ST Funding(s)	13.4609	12.8258
13	Cap Funds/Liabilities(L)	11.4537	10.4268
14	Sub Debt/Cap Funds	7.5239	13.3812
15	Net Interest Margin	4.1845	4.5305
16	Net Interest Revenues/Avg Assets	3.6280	4.0294
17	Oth Op Inc/Avg Asset	3.8448	2.1366
18	Non Interest Expenses/Avg Assets	5.6982	4.0304
19	Pre-Tax Op Inc/Avg Assets	1.8890	2.0377
20	Non Op Items & Tax/Avg Assets	−0.3973	−0.1795
21	Return on Average Assets(ROAA)	1.3963	1.9431
22	Return on average Equity (ROAE)	14.3225	15.4577
23	Non Op Items/Net Income	−5.4292	3.8968
24	Cost to Income Ratio	57.3107	55.7193
25	Recurring Earning Power	2.4868	2.9465

Table 6.1   *(continued)*

#	Indicator	Mean for non-investment grade banks (Status 0)	Mean for investment grade banks (Status 1)
26	Net Loans/Total Assets	40.4006	58.3013
27	Net Loans/Customer & ST Funding	57.9011	125.1756
28	Net Loans/Tot Dep & Bor	49.5629	76.9383
29	Liquidity Asset/Cust & ST Funding	34.2381	44.9915
30	Liquidity Asset/Tot Dep & Bor	29.6734	16.6503

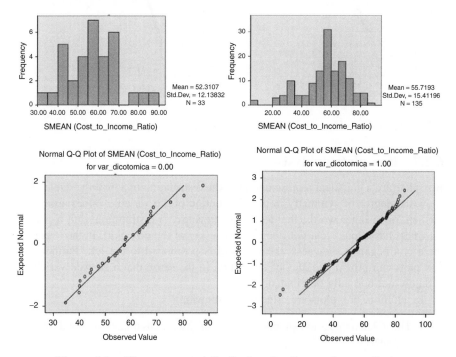

Figure 6.1   Histograms and Q–Q plots for Cost to Income Ratio.

Overall, half of the 30 ratios in the sample have a split judgment in the two subsamples, whereas the other half are normal or non-normal in both subsamples. For an analytical ratio by ratio indication of normality based on graphs, refer to the summary in Table 6.10.

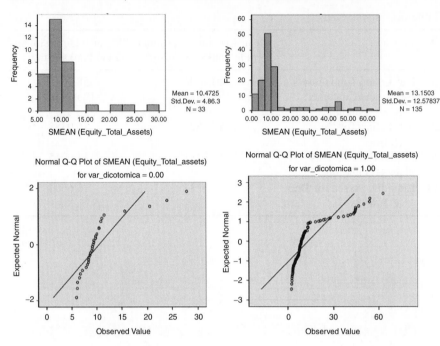

Figure 6.2   Histograms and Q–Q plots for Equity / Total Asset ratio.

6.2.7   Box plots analysis

Box plots are a graphical representation of the distribution of a given variable. They show the degree of symmetry in the distribution of the variable values around the median, or in other words, the degree of skewness: a distribution with a positive skewness has a long right tail (the upper whisker of the box plot). In the event that there is a negative skewness, the long tail is positioned to the left. Furthermore, box plots can provide an idea of normality, in the sense that a skewed variable is certainly not normally distributed.

In addition, box plots are effective in showing:

- The relative position of medians for the two subsamples of good and bad borrowers.

- The existence of outliers and extremes values.

- If the two distributions of good and bad observations are fairly separated or, conversely, largely overlapped; in effect, the box represents the inter-quartile range (50% of the observations in the center of the distribution) and the whiskers represent the extension of the distribution of the remaining 25% of observations in the two tails. The larger the overlap of the box and whiskers of the good and bad subsamples, the lower the power of the examined variable to distinguish between non-investment grade and investment grade banks.

More than two-thirds of the ratios considered have largely overlapped distributions of values in the two subsamples. Conversely, eight ratios present a much lower overlapping area and suggest a higher discriminatory power of indicators.

For instance, consider the Loan Loss Reserves / Gross Loans box plot in Figure 6.3. Values for non-investment grade banks are much higher than those of the investment grade banks. The circumstances are completely different for Net Loans / Total Assets ratio: in Figure 6.4 the overlapping area involves both the boxes and whiskers of the two distributions.

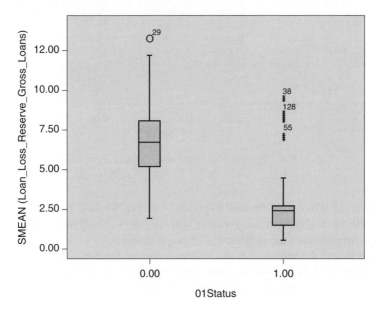

Figure 6.3   Box-plots for Loan Loss Reserves / Gross Loans.

Now, consider the symmetry of distributions depicted by the four box plots examined as examples. Only the ratio Loan Loss Reserves / Gross Loans for the non-investment grade banks clearly has the median at approximately the middle of the graph and two whiskers of about the same length. Symmetry has to be certainly excluded for the same indicator when considering investment grade banks and only a good degree of benevolence can induce the consideration of the distributions for Net Loans / Total Assets ratio as symmetric.

6.2.8   Normality tests

The Kolmogorov–Smirnov and Saphiro–Wilk tests are used to assess the normality of distributions; they compare sample values with a normal distribution which have the same mean and standard deviation. The null hypothesis is that there is no difference between the two distributions. So, we look for 'non-significant tests' in

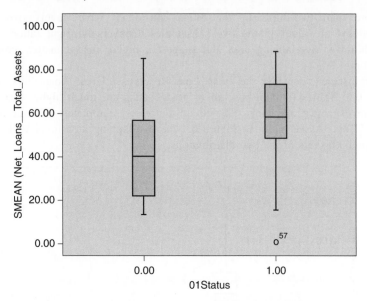

Figure 6.4   Box-plot of Net Loans / Total Assets ratio.

order to avoid rejection of the null hypothesis: that is, we look for p-values higher than the level of significance (α-value), usually set at 5%. Using normality tests allows us to eliminate the subjectivity of graph analysis.

According to the statistical tests mentioned above, about two-thirds of the ratios present normal distributions for both good and bad cases. Other ratios present significant tests for non-investment grade subsamples and, therefore, non-normal distributions; they are:

- Loan loss reserve / Gross loans.

- Loan loss provisions / Net interest revenues.

- Impaired loans / Gross loans (only the Saphiro–Wilk test is significant).

- Net interest margin.

- Net interest revenue / Average assets.

- ROAE.

- Cost to income ratio.

- Net loans / Total assets.

- Net loans / Total deposits and borrowing (only the Kolmogorov–Smirnov test is significant).

6.2.9   Homogeneity of variance tests

The homogeneity of variance in the two subsamples of good and bad cases is another property often required by multivariate statistical techniques. Levine's test is used to verify this property of data distribution. It works in a similar way to the Kolmogorov–Smirnov and Saphiro–Wilk tests of normality. The null hypothesis is that variances are homogenous, consequently, we look for 'non significant' tests (p-value higher than 5%) in order to avoid rejection of the null hypothesis. As indicated in Table 6.2, 50% of the ratios present homogeneity of variance.

Table 6.2   Levine's tests p-values.

#	Indicator	p-values
1	Loan_Loss_Reserve__Gross_Loans	0.013
2	Loan_Loss_Prov__Net_Int_Rev	0.770
3	Loan_Loss_Res__Impaired_Loans	0.003
4	Impaired_Loans__Gross_Loans	0.002
5	Total_Capital_Ratio	0.326
6	Equity__Total_Assets	0.002
7	Equity__Net_Loans_	0.002
8	Equity__Dep__ST_Funding	0.000
9	Equity__Liabilities	0.001
10	Cap_Funds__Tot_Assets	0.508
11	Cap_Funds__Net_Loans	0.000
12	Cap_Funds__Dep__ST_Funding	0.730
13	Capital_Funds__Liabilities	0.632
14	Subord_Debt__Cap_Funds	0.072
15	Net_Interest_Margin	0.016
16	Net_Int_Rev__Avg_Assets	0.013
17	Oth_Op_Inc__Avg_Assets	0.030
18	Non_Int_Exp__Avg_Assets	0.296
19	PreTax_Op_Inc__Avg_Assets	0.045
20	Non_Op_Items__Taxes__Avg_Ast	0.233
21	Return_on_Average_Assets_ROAA	0.041
22	Return_on_Average_Equity_ROAE	0.267
23	Non_Op_Items__Net_Income	0.303
24	Cost_to_Income_Ratio	0.253
25	Recurring_Earning_Power	0.004
26	Net_Loans__Total_Assets	0.712
27	Net_Loans__Customer__ST_Funding	0.016
28	Net_Loans__Tot_Dep__Bor	0.290
29	Liquid_Assets__Cust__ST_Funding	0.113
30	Liquid_Assets__Tot_Dep__Bor	0.872

6.2.10   F-ratio and F-Test

The power of an indicator to discriminate between good and bad borrowers is important in order to qualify the indicator as being useful for building a statistical based rating system. Discriminant power can be assessed using the F-test; this test can control the significance in the difference between the means of good and bad cases. The null hypothesis here is that there is no difference between the two means. So, we hope to reject the null hypothesis by having 'significant' tests, that is to say, p-values lower than the required α-value (the usual 5%). Table 6.3 shows that many ratios do not have discriminant power according to the F-test. This test also produces an F-value for each ratio; it can be considered an indicator of discriminatory power. The most powerful ratio is Capital Funds / Net Loans; it has a p-value of approximately zero and an F-value of 135.

6.2.11   ROC curves

In addition to the F-test, ratios' discriminatory power can be assessed by using ROC curves and its related measure AuROC (area under the ROC curve). In the undergoing analysis, the 'positive actual state' is the value 0 of the dichotomous variable (non-investment grade). To build ROC curves it is necessary to sort the observations in descending order, starting with the maximum value of a ratio expressing the positive actual state to the minimum value of the ratio. Subsequently, it is necessary to divide indicators into two groups:

1. Group 'SMALLER', where 'smaller values of the test result variable(s) indicate stronger evidence for a positive actual state' (Table 6.4).

2. Group 'LARGER': where 'larger values of the test result variable(s) indicate stronger evidence for a positive actual state' (Table 6.5).

6.2.12   Correlations

To analyze the correlation, both Pearson's Correlation and Spearman's Correlation can be used. Table 6.10 summarizes the results from the Spearman's correlation matrix:

- YES indicates a correlation higher than 0.7 with at least another variable;

- YES* indicates a correlation higher than 0.8 with at least another variable;

- NO indicates a correlation not higher than 0.7 with all other variables.

6.2.13   Outliers

Box plots and Q–Q plots have outlined the existence of outliers. However, in regards to non-investment grade observations, the number of outliers goes from a

Table 6.3   F-values and p-values of F-test.

#	Indicator	F-value	p-value
1	Loan_Loss_Reserve__Gross_Loans	93.432	0.000
2	Loan_Loss_Prov__Net_Int_Rev	5.977	0.016
3	Loan_Loss_Res__Impaired_Loans	13.493	0.000
4	Impaired_Loans__Gross_Loans	80.055	0.000
5	Total_Capital_Ratio	0.029	0.866
6	Equity__Total_Assets	1.438	0.232
7	Equity__Net_Loans_	25.963	0.000
8	Equity__Dep__ST_Funding	4.244	0.041
9	Equity__Liabilities	2.387	0.124
10	Cap_Funds__Tot_Assets	2.836	0.094
11	Cap_Funds__Net_Loans	135.788	0.000
12	Cap_Funds__Dep__ST_Funding	0.633	0.428
13	Capital_Funds__Liabilities	2.695	0.103
14	Subord_Debt__Cap_Funds	12.815	0.000
15	Net_Interest_Margin	0.127	0.722
16	Net_Int_Rev__Avg_Assets	0.217	0.642
17	Oth_Op_Inc__Avg_Assets	3.177	0.077
18	Non_Int_Exp__Avg_Assets	3.081	0.081
19	PreTax_Op_Inc__Avg_Assets	0.304	0.582
20	Non_Op_Items__Taxes__Avg_Ast	0.267	0.606
21	Return_on_Average_Assets_ROAA	1.273	0.261
22	Return_on_Average_Equity_ROAE	0.422	0.517
23	Non_Op_Items__Net_Income	4.978	0.027
24	Cost_to_Income_Ratio	0.305	0.581
25	Recurring_Earning_Power	0.572	0.451
26	Net_Loans__Total_Assets	20.255	0.000
27	Net_Loans__Customer__ST_Funding	8.829	0.003
28	Net_Loans__Tot_Dep__Bor	23.144	0.000
29	Liquid_Assets__Cust__ST_Funding	0.217	0.642
30	Liquid_Assets__Tot_Dep__Bor	3.527	0.062

Table 6.4   Area under the curve (Group Smaller).

Test result variable(s)	AuROC
SMEAN(Loan_Loss_Res__Impaired_Loans)	0.740
SMEAN(Equity__Dep__ST_Funding)	0.528
SMEAN(Non_Op_Items__Taxes__Avg_Ast)	0.509
SMEAN(Non_Op_Items__Net_Income)	0.714
SMEAN(Net_Loans__Total_Assets)	0.751
SMEAN(Net_Loans__Customer__ST_Funding)	0.776
SMEAN(Net_Loans__Tot_Dep__Bor)	0.813
SMEAN(Subord_Debt__Cap_Funds)	0.734
SMEAN(Return_on_Average_Equity_ROAE)	0.511

Table 6.5   Area under the curve (group larger).

Test result variable(s)	AuROC
SMEAN(Loan_Loss_Reserve__Gross_Loans)	0.898
SMEAN(Loan_Loss_Prov__Net_Int_Rev)	0.733
SMEAN(Impaired_Loans__Gross_Loans)	0.855
SMEAN(Total_Capital_Ratio)	0.590
SMEAN(Equity__Total_Assets)	0.529
SMEAN(Equity__Net_Loans)	0.785
SMEAN(Equity__Liabilities)	0.526
SMEAN(Cap_Funds__Tot_Assets)	0.642
SMEAN(Cap_Funds__Net_Loans)	0.925
SMEAN(Cap_Funds__Dep__ST_Funding)	0.613
SMEAN(Capital_Funds__Liabilities)	0.635

minimum of none to a maximum of five for each indicator, whereas for investment grade observations, the number of outliers ranges between one and 22; in particular, there are:

- three indicators with 22 outliers;

- one indicator with 17 outliers;

- two indicators with16 outliers;

- three indicators with14 outliers;

- two indicators with13 outliers;

We have decided to exclude ratios with more than 15 outliers; that is, three indicators with 22 outliers (Equity/Total Assets, Equity/Deposit and Short Term Funding, Equity/Liabilities), one indicator with 17 outliers (ROAA), and two indicators with 16 outliers (Recurring Earning Power and Non-Interest Expense/Average Asset). For other ratios, it has been assumed that the number of outliers is not high enough to require any management.

6.2.14   Short listing and linear discriminant analysis

Univariate and bivariate analysis produce the short list of indicators considered suitable for subsequent steps. Out of 30 indicators originally available in the dataset, 15 ratios have been entered into the short list because it is more consistent with the required properties and/or more powerful to discriminate. The ratios' characteristics that are considered critical to allow ratios to be short listed are:

- Discriminant power in terms of AuROC.

- Number of outliers.

- Monotonicity.

- Significant F-test and F-ratio value.

- Normality and homogeneity of variance.

Linear discriminant analysis has been applied with a stepwise methodology for ratios selection. From the 15 ratios in the short list, eight have been selected by the procedure. It is worth noting that none of the ratios excluded from the short list would have been entered into the discriminant function if the stepwise procedure had been applied to the whole set of 30 ratios. This notion confirms the accuracy of the short listing phase.

Table 6.6 and Table 6.7 report coefficients of the estimated discriminant function (canonical coefficients and structure matrix coefficients). The resulting contingency matrix is reported in Table 6.8. More than 98% of the observations are correctly classified. Figure 6.5 shows the ROC curve, whereas Table 6.9 reports the AuROC value. Results appear to be satisfactory.

Table 6.6   Canonical discriminant function coefficients.

Variable	Coefficient value
SMEAN(Loan_Loss_Reserve__Gross_Loans)	0.606
SMEAN(Equity__Net_Loans)	−0.057
SMEAN(Cap_Funds__Tot_Assets)	1.790
SMEAN(Cap_Funds__Net_Loans)	0.144
SMEAN(Capital_Funds__Liabilities)	−1.505
SMEAN(Subord_Debt__Cap_Funds)	−0.034
SMEAN(Non_Op_Items__Net_Income)	−0.010
SMEAN(Net_Loans__Customer__ST_Funding)	−0.004
(Constant)	−3.903

Table 6.7   Structure matrix.

Variable	Coefficient value
SMEAN(Cap_Funds__Net_Loans)	0.518
SMEAN(Loan_Loss_Reserve__Gross_Loans)	0.429
SMEAN(Impaired_Loans__Gross_Loans)	0.374
SMEAN(Equity__Net Loans)	0.226
SMEAN(Net_Loans__Total_Assets)	−0.224
SMEAN(Subord_Debt__Cap_Funds)	−0.159
SMEAN(Net_Loans__Tot_Dep__Bor)	−0.148
SMEAN(Net_Loans__Customer__ST_Funding)	−0.132
SMEAN(Liquid_Assets__Tot_Dep__Bor)	0.110
SMEAN(Non_Op_Items__Net_Income)	−0.099
SMEAN(Cap_Funds__Tot_Assets)	0.075
SMEAN(Capital_Funds__Liabilities)	0.073
SMEAN(Loan_Loss_Prov__Net_Int_Rev)	−0.034
SMEAN(Loan_Loss_Res__Impaired_Loans)	−0.031
SMEAN(Oth_Op_Inc__Avg_Assets)	0.018

Table 6.8   Contingency Table.

			Classification results[a]		
		01Status	Predicted group membership		
			non investment grade	investment grade	Total
Original	Count	non investment grade	30	3	33
		investment grade	0	135	135
	%	non investment grade	90.9	9.1	100.0
		investment grade	0.0	100.0	100.0

[a]98.2% of original grouped cases correctly classified.

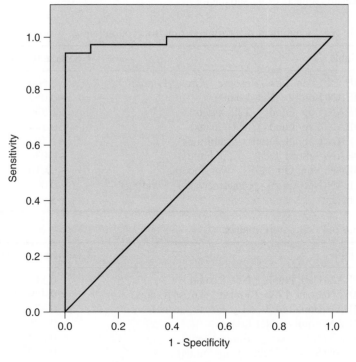

Figure 6.5   ROC curve.

6.3   The 'Validation report' by the validation unit

From here on, the case study presents the main criticisms of the 'Development Report' that the validation unit should outline. They will be presented step by step in relation to the structure of the Development Report.

Table 6.9   Area Under the Curve.

Test result variable(s): Discriminant scores from function 1 for analysis 1		
	Asymptotic 95% confidence interval	
Area	Lower bound	Upper bound
0.986	0.963	1.009

1. The model development process and the dataset

Overall, the structure of the Development Report is adequate, but its content is largely incomplete.

Firstly, the precise perimeter of application of the model under development, PanAlp Bank's loan portfolio structure, and its relationship with the dataset are not clear.

Secondly, two steps have been completely missed: (i) calibration of the model in order to achieve rating classes and associated cardinal measures of risk; (ii) indications of model limitations and the possible room for integrating qualitative/judgmental information, as well as other quantitative information not included into the dataset. No insights have been given concerning either why information (such as ownership structure, issued securities and their prices) have not been used to develop the model even if present in the dataset or how it is supposed to utilize them.

Thirdly, the use of a shadow rating approach could have merited a clarification on its limitations and a statement of precautions to be used. In effect, this approach determines model building and evaluation based on the comparisons of two predictions: the model's predictions against rating agencies' predictions (rather than model's predictions against facts, such as defaults). It is simply stated that the methodology is justified by the lack of a sufficient amount of defaults in the banking industry. In reality, this is not true of banks in general, but it may be true in terms of types of banks borrowing from PanAlp Bank (the report lacks any indication of the nature of PanAlp Bank's loan portfolio).

From examining the issues relating to the dataset, the following observations are crucial:

• The sample is, as a whole, too small (168 observations, 33 non-investment grade and 135 investment grade banks). The proportion of bad to good cases is sufficient, so it has avoided one of the typical issues of default-based datasets (too low percentage of defaults).

• It is unclear in terms of which selection criteria has been used to extract the 168 observations from the thousands of banks recorded in Bankscope, and it is also equally unclear which relation they have with the PanAlp Bank loan portfolio. Randomly selecting such a small sample from the Bankscope universe could be a risky solution for many reasons:

(a) banks with different specialization (investment banks, commercial banks, merchant banks) have completely different structures of accounting data, independent from being investment or speculative grade;

(b) banks of differing sizes usually have varying balance sheet and income statement structures;

(c) banks of different countries may have specificities determined by local regulations and market conditions.

There are many regulatory requirements set by Basel II regarding data to be used for setting rating assignment processes, performing rating quantification, and checking predictive power. For instance, the model must be accurate on average across the range of borrowers or facilities to which the bank is exposed and there must be no known material biases (Basel Committee, §417). A bank must demonstrate in its analysis that the estimates are reflecting underwriting standards (§462). Irrespective of whether a bank is using external, internal, or pooled data sources, or a combination of the three, for its PD estimation, the length of the underlying historical observation period used must be at least five years for at least one source (§463). Eventually, using agencies' ratings as a proxy of risk will require that, during the rating quantification process, the default definition set by Basel II must be related to agencies' default definitions (§462: The bank's analysis must include a comparison of the default definitions used, subject to the requirements in Paragraphs 452 to 457).

- A great weakness of the dataset is that it contains three observations for each of the 56 banks considered. This means that observations are not truly independent.

- A further important limitation of the dataset is that its small size has not allowed any split between a development sample and a validation sample. The consequence is that there has not been any hold-out validation of the model's performance.

- Finally, it is not clear which rules have been used in the case of 'split ratings' leading to a different collocation of the bank in terms of investment or non-investment grades (in case more than one rating agency is rating a borrower and their ratings are different, it is commonly said that there are 'split ratings').

2. Missing value analysis

Data cleansing is an important first step to be carried out after collecting data. It is at least necessary to verify duplicate cases and missing values. The former analysis has not been reported. In any case, the limited size of the sample should exclude the presence of undetected duplicates coming from the data collection process. The latter analysis is poorly reported, because nothing has been mentioned in regards to the number and concentration of missing values per ratio and per observation: these are considered key information, and are needed in order to make and validate decisions concerning alternative missing value management approaches. In addition, the means of the two subsamples have been substituted in place of the missing values (investment grade and non-investment grade). This is one of the more traditional alternatives in managing missing values when they are not too concentrated in specific ratios; and for the analysis sample, it is generally considered acceptable to use the two subsample means rather than the overall mean. Here the point is that means can be strongly affected by outliers (whose analysis has not yet be done); this is why it is advisable to use the median instead.

3. Economic meaning of ratios

Ratios included into the dataset concern loan portfolio quality (1–4), bank's capitalization and financial strength (5–14), bank's profitability (15–25), and maturity mismatch and liquidity (25–30). Therefore, they cover a wide range of information. However, no reason has been given in the development report as to why 30 have been selected out of the 36 ratios proposed by Bankscope. In addition, many ratios such as leverage and profitability ratios are sensible to banks' size and specialization.

Some ratios, such as Non-Operating Items and Taxes/Average Assets, do not have a concise theoretical economic relation with the default, and can be immediately excluded from the list for this reason.

4. Monotonicity

An indicator such as 'Non-operating items / Net income' is not structurally monotonic, as the denominator (and also the numerator) can change sign, inverting the relationship with the probability of default. A second verification on monotonicity regards the empirical monotonicity. It is not clear from the Development Report if both forms of monotonicity have been assessed and which technical approach has been used to verify empirical monotonicity.

5. Means position versus working hypothesis

For each ratio, a working hypothesis concerning the sign of the ratio's relationship to the probability of default has been set in the Development Report. Subsequently, the working hypothesis has been checked on the basis of the relative position of the means of the two subsamples. Approximately one-third of the ratios do not satisfy expectations. It may well have been useful to also have a check based on medians or on trimmed means; very often these checks are more successful and indicate that there is an alarming presence of outliers. In general, descriptive statistics concerning ratios have been missed in the report.

6. Normality of distributions: Histograms and Normal Q–Q plots.

Histograms of 'Cost to Income' ratio show a normal distribution for investment grade banks subsamples, but not for the other subsample (it is not 'peaked around the mean'). A subjective view of Q–Q plots for the same ratio leads to consider normal also the latter distribution (points are close to the straight line). For the indicator Equity/Total Asset both graphs clearly show a non-normal distribution. The subjectivity of these analyses is apparent.

7. Distributions analysis using box plots

Box plots have only been used to check if the good and bad subsamples distributions' are sufficiently separated. Other information could have been acquired from box plots and transferred into the Development Report. In addition, separation analysis based on box plots is very subjective. For instance, Net Loans / Total Asset ratio is said not to have discriminatory power due to the large overlapping of the two distributions. On the contrary, notwithstanding some overlapping, there is an acceptable value of the F statistic and a p-value of the F-test lower than the 0.05 criterion (to reject the null hypothesis). Consequently, if the decision to exclude this variable from the short list (that is deducible by the Structure Matrix where all short listed ratios are accounted for) was based on the box plot analysis, this exclusion is not advisable.

Eventually, Wilks' Lambda statistic could have been calculated, as it is also useful in univariate analysis, to assess the discriminatory power of a financial ratio.

8. Normality test

9. Homogeneity of variance test

Information provided on normality tests in the Development Report can only be seen in the Summary Table, expressed as 'Yes' or 'No'. Note that a much larger number of ratios is considered to be normally distributed according to the test rather than according to the judgmental analysis based on histograms and Q–Q plots. This is certainly due to the wide subjectivity of the Q–Q plot analysis.

Levine's test results are detailed with their p-values in the Development Report (Table 6.2). The null hypothesis is that the variance of the bad and good subsamples is equal. Therefore, the hope is not to reject the null hypothesis, that is, to have p-value greater than 0.05. Therefore, what has been reported in the Summary Table 6.10 is contradictory to what is stated in Table 6.2.

10. F-ratio and F-Test

11. ROC Curves

The F-ratio and F-test evaluate the variability explained by 'the model' compared to the unexplained variability. According to the F-ratio, the best

Table 6.10 Summary table of indicators economic and statistical propertie.

# Ratio (%)	It has economic meaning (Y/N); sign of relation with PD (+,-)	Monotonicity	Means are aligned with hypothesis	Histograms and Q–Q plot suggest normality of distributions	Box plots suggest separation of distributions	Normality test (K-S; SW) is not significant (p-value > 0.05)	Homogeneity of variance (Levine's) test is not significant (p-value > 0.05)	F-test is significant (p-value < 0.05)	F-value	AuROC (%)	High Spearman's correlation	Missing and outliers
Legend	Y/N+–	Yes/No	Y/N	0 = Y/N, 1 = Y/N	Yes/No	Y/N	Y/N	Y/N	Value	Value	Yes/Yes*/No	M/O
1 Loan Loss Reserve/ Gross Loans	Y–	Yes	N	YN	Yes	NY	Y	Y	93.432	89.8	Yes*	M/O
2 Loan Loss Prov/ Net Int Rev	Y–	Yes	N	YN	No	NY	N	Y	5.977	73.3	No	M/O
3 Loan Loss Reserve/ Impair ed Loans	Y–	Yes	Y	YN	Yes	YY	Y	Y	13.493	74.0	No	M/O
4 Impaired Loans/Gross Loans	Y+	Yes	Y	YN	Yes	YN	Y	Y	80.055	85.5	Yes*	M/O
5 Total Capital Ratio	Y–	Yes	Y	YN	No	YY	N	N	0.029	59.0	No	M/O
6 Equity/Tot Assets	Y–	Yes	Y	NN	No	YY	Y	N	1.438	52.9	Yes*	M/O
7 Equity/Net Loans	Y–	Yes	N	NN	Yes	YY	Y	Y	25.963	78.5	Yes	M/O
8 Equity/Dep & ST Funding	Y–	Yes	N	NN	No	YY	Y	Y	4.244	52.8	Yes*	M/O

(continued)

Table 6.10 (continued)

# Ratio (%)	It has economic meaning (Y/N); sign of relation with PD (+,-)	Monotonicity	Means are aligned with hypothesis	Histograms and Q–Q plot suggest normality of distributions	Box plots suggest separation of distributions	Normality test (K-S; SW) is not significant (p-value > 0.05)	Homogeneity of variance (Levine's) test is not significant (p-value > 0.05)	F-test is significant (p-value < 0.05)	F-value	AuROC (%)	High Spearman's correlation	Missing and outliers
Legend	Y/N+−	Yes/No	Y/N	0 = Y/N 1 = Y/N	Yes/No	Y/N	Y/N	Y/N	Value	Value	Yes/Yes*/No	M/O
9 Equity/Liabilities	Y −	Yes	Y	NN	No	YY	Y	N	2.387	52.6	Yes*	M/O
10 Cap Funds/Tot Assets	Y −	Yes	N	NY	No	YY	N	N	2.836	64.2	Yes*	M/O
11 Cap Funds/Net Loans	Y −	Yes	Y	NN	Yes	YY	Y	Y	135.788	92.5	Yes	M/O
12 Cap Funds/Dep & ST Funding	Y −	Yes	N	NY	No	YY	N	N	0.633	61.3	Yes	M/O
13 Cap Funds/Liabilities	Y −	Yes	N	NY	No	YY	N	N	2.695	63.5	Yes*	M/O
14 Sub Debt/Cap Funds	Y +	Yes	N	NY	Yes	YY	N	Y	12.815	73.4	Yes*	M/O
15 Net Interest Margin	Y −	No	Y	YN	No	NY	Y	N	0.127	64.4	Yes*	M/O
16 Net Int Rev/Avg Assets	Y −	Yes	Y	YN	No	NY	Y	N	0.217	64.2	Yes*	M/O
17 Oth Op Inc/Avg Asset	Y −	Yes	N	NN	No	YY	Y	N	3.177	63.4	Yes	M/O
18 Non Int Exp/Avg Assets	Y +	Yes	Y	NN	No	YY	N	N	3.081	65.0	Yes	M/O

19	Pre-Tax Op Inc/Avg Assets	Y –	Yes	Y	NN	No	YY	Y	N	0.034	50.5	Yes*	M/O
20	Non Op Items & Tax/Avg Assets	Y –	Yes	Y	NN	No	YY	N	N	0.267	50.9	Yes	M/O
21	Return on Average Assets(ROAA)	Y –	Yes	Y	YN	No	YY	Y	N	1.273	50.5	Yes	M/O
22	Return on average Equity(ROAE)	Y –	Yes	Y	YY	No	NY	N	N	0.422	51.1	Yes	M/O
23	Non Op Items/Net Income	Y +	Yes	Y	NN	No	YY	N	Y	4.978	71.4	No	M/O
24	Cost to Income Ratio	Y +	Yes	Y	YY	No	NY	N	N	0.305	51.3	No	M/O
25	Recurring Earning Power	Y –	Yes	Y	YN	No	YY	Y	N	0.572	62.1	Yes*	M/O
26	Net Loans/Tot Assets	Y +	Yes	N	YY	No	NY	N	Y	20.255	75.1	Yes*	M/O
27	Net Loans/Customer & ST Funding	Y +	Yes	Y	YN	Yes	YY	Y	Y	8.829	77.6	Yes	M/O
28	Net Loans/Tot Dep & Bor	Y +	Yes	N	NY	Yes	NY	N	Y	23.144	81.3	Yes*	M/O
29	Liquidity Assets/Cust & ST Funding	Y –	Yes	Y	NN	No	YY	N	N	0.217	67.9	Yes	M/O
30	Liquidity Assets/Tot Dep & Bor	Y –	Yes	N	NN	No	YY	N	N	3.527	78.6	Yes	M/O

indicator is Capital Funds/Net Loans (p-value is close to zero and F-ratio is the highest value). Note that for many ratios, including the Total Capital Ratio, the F-test is not significant.

Only some of the 30 ratios have their AuROC reported in the two tables detailing AuROC values.

12. Correlations

It would be better to add the traditional Pearson's correlation to the non-parametric correlation (Spearman) analysis, rather than reporting only the latter. No motive has been given for the decision in the Development Report. Also, the choice of a cut-off, defining the correlation unacceptable, set at 70% is not motivated. Statistical tests for correlation measures are missed. The concise reporting of correlation phenomena does not allow an understanding of which pairs of ratios are correlated.

13. Outliers

Firstly, no definition of 'outlier' has been given. A definition is necessary because the concept of outliers is not univocally defined in statistics. The treatment of outliers chosen by the development unit does not tend to smooth the problem: ratios were neither treated nor cancelled from the short list. There are no indications of having evaluated other solutions, such as appropriate logistic transformation of variables. While four excluded variables have low AuROC values, two of them have AuROC values higher than 60%. These variables are those with the minimum number of outliers (16) above the cut-off limit (set at 15 cases): the decision to set the limit at 15 cases is therefore important but has not been supported by any argument in the Development Report.

14. Short listing and Discriminant Function estimation

This chapter of the Development Report is particularly weak. Details regarding the rules for short listing ratios have not been adequately defined. No explanation has been given as to why discriminant analysis, rather than other statistical techniques has been chosen to build the model.

Single steps of the stepwise procedure have not been reported, nor have the entering and exiting criteria.

The Development Report does not address an important issue of the estimated function: two ratios (Equity/Net Loans and Capital Funds/Liabilities) have conflicting signs of their coefficients in the canonical discriminant function and in the structure matrix. Perhaps correlation phenomena among variables are at the basis of this worrying signal. We have already stated that correlation has been poorly reported and perhaps insufficiently examined by the development unit. We can now add that a correlation matrix among the variables entered into the model would have been beneficial.

If coefficients' signs are different in the structure matrix and in the canonical discriminant function, the coherence check of coefficients signs with working hypotheses (set at the outset of univariate analysis) would also be in trouble. The check is also made difficult because centroids values of good and bad cases are not indicated; as a result, the direction (risk or creditworthiness) of the discriminant function and of scores is not clear. If we assume that bad banks centroid is higher than that of good banks, five out of eight ratios entered into the discriminant function have canonical coefficients not aligned with working hypotheses (Loan Loss Reserve/Gross Loans, Capital funds/Total assets, Capital Funds/Net loans, Subordinated Debt/Capital Fund, and Net Loans/Customer&Short-Term funding).

Standardized canonical discriminant function coefficients are not reported. The combined analysis of these coefficients and those of the structure matrix can give useful insights on the importance of the contribution of each ratio in the model.

In the Development Report it was affirmed that all ratios in the dataset were monotonic with the exception of Non-operating items/Net income. The problem is that this indicator has been entered into the discriminant function.

The contingency table shows a good capacity of the model to the overall reclassification of good and bad cases, with a hit rate of 98.2%. The 'Error I' rate (the most expensive for the bank) is approximately 9%. There is no information regarding the chosen cut-off of scores in order to define predicted good and bad cases.

ROC curves calculated on scores are very promising: the AuROC is at an excellent 98.6%. However, this is an in-the-sample performance; no attempt has been made to validate the model out-of-sample and perhaps also out-of-time and out-of-universe. In addition, the leave-one-out classification which gives an artificial hold-out performance is lacking. Only the confidence interval for the AuROC is reported.

Other useful statistical indicators of the discriminant function performance (such as ETA squared, M test and Wilks' Lambda) have been missed.

No attempts have been made to define classes of scores in order to arrive at ratings and to verify the degree of overlapping of confidence intervals of the default probability associated with ratings.

7

Ratings usage opportunities and warnings

7.1 Internal ratings: critical to credit risk management

The current quantitative techniques for managing credit risk are greatly innovating competition in financial markets. In bank management, all four areas of credit-related activities listed in Table 7.1 are radically changing. Directly or indirectly, internal rating systems are critical in all applications. In addition, probabilities of default and loss given default represent the necessary measures to other more complex risk measurements (which can be read downwards in Figure 7.1, towards mark-to-market measures, and rightwards, towards measures of unexpected events). Note that nowadays, due to the 2008 financial crisis, we should add liquidity risk and relative tools such as stress tests to Figure 7.1; in fact, today liquidity risk is recognized as an important source of credit risk.

Rating systems are not only critical to modern credit risk management but – different from other tools – they are also largely decentralized in banks' organizations. Rating processes involve loan and credit officers in remote bank branches as well as all superior decisional levels up to the board of directors. As they are essential to competitiveness and in creating economic value, on the one hand, they have an impact on shareholders and, on the other hand, on customers as well. In particular, internal rating systems profoundly change the relationship between banks and firms.

From this perspective, it should be noted that banks create value for their borrowers not only by lending them money *per se*, but also by assisting and monitoring

Developing, Validating and Using Internal Ratings: Methodologies and Case Studies   Giacomo De Laurentis,
Renato Maino and Luca Molteni   © 2010 John Wiley & Sons, Ltd

Table 7.1   Credit risk management applications.

1. process auditing		
2. portfolio reporting		
3. credit management	3.1 single credit decisions	3.1.a) credit culture
		3.1.b) administrative actions
		3.1.c) credit limits
		3.1.d) risk adjusted profitability
		3.1.e) employee compensation
	3.2 portfolio management	3.2.a) capital allocation
		3.2.b) portfolio optimization by secondary and derivative markets
		3.2.c) overall bank risk management
4. credit administration	4.1 provisioning	
	4.2 regulatory compliance	
	4.3 transparency	

them through all the stages of their relationship (Figure 7.2). The relationship takes the form of an articulated process, whose central components are: creditworthiness assessment, daily monitoring and early detection of difficulty signals, and eventual credit workouts through legal means. These steps are dispersed between head offices and branches. At the two extremes of this process there are two important areas that strongly contribute to characterize bank–firm relationships: on one side, the vision and mission of bank credit activity and, on the other side, the portfolio management of credit exposures.

Rating systems change the vision of bank lending, spreading available strategic options for:

- setting banks' organizational structures for dealing with customers,

- combining credit risk management within the general risk management of a bank,

- participating to banks profit generation.

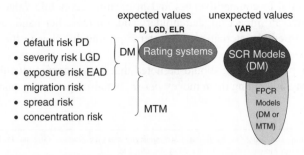

Figure 7.1   Credit risk profiles, tools, and measures.

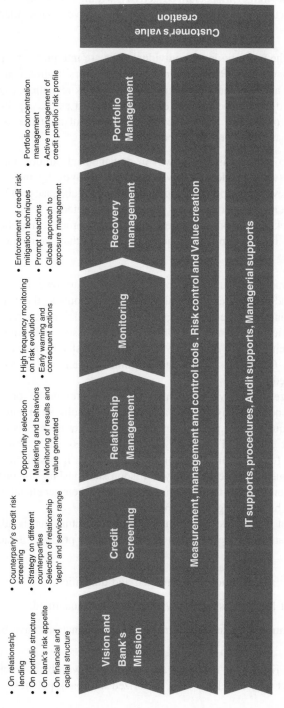

Figure 7.2   Value chain of bank lending.

Lending is no longer a specialized activity within banks, with its own rules, language, and culture. Instead, lending fully involves bank management at two levels:

- One is at the portfolio level: individual risks are interpreted within a general framework of risks. Risk concentration and diversification can be managed in order to achieve important gains in terms of risk reduction, capital savings, leverage effects, and earnings generation, as well as the capability to assist single borrowers and sectors beyond the stringent limit that the traditional 'hold to maturity approach' imposes.

- At the second level is the growth in management culture that takes place internally (using models to drive decisions, analytically measuring results, rationally evaluating policies and business model adequacy, linking compensation schemes to performance properly interpreted in terms of risk-return for the bank, and so on) and externally (setting commercial policies in a way that risk is related to price and quantity of credit that can be underwritten by the bank).

In addition, more efficient screening processes lead to lower cost of risk. Early warning tools and procedures reduce the cost of work-outs as they limit the exposure at default and increase the safeguard of collaterals and guarantees. A good portfolio diversification releases the constraints of excessive concentration that new lending opportunities could otherwise face. Non-interest earnings can be added to interest margins. Capital absorption can be minimized and risk adjusted return on capital maximized.

To summarize, pro-active credit management creates profit for the bank and value for the economy. Many of these achievements require an organization clearly geared towards quantitative risk management and are not even possible with more traditional credit cultures. This is why internal rating systems should not merely be seen as tools for credit underwriting decisions. Even if a bank introduces them only for this immediate objective, they will soon start pushing for further changes in the organization in order to allow the spreading of their beneficial effects within the entire set of bank management processes.

One signal of the extensive infiltration of rating culture that is taking place in daily bank operations is the adoption of risk-adjusted pricing. In fact, it implies that:

- borrower's risk is rationally examined and quantified,

- severity of loss risk is rationally examined and quantified,

- risk mitigation alternatives and risk/return combinations have been checked,

- individual risk-taking decisions are coordinated with bank credit policies and market pressures have been factorized into bank policies.

On the contrary, Bank Old, which does not develop risk-adjusted pricing procedures exposes itself to adverse selection when competing with Bank Alpha that has developed a simple three-class rating system and has started to differentiate prices (Figure 7.3). Bad borrowers find it cheaper to pay 8% to Bank Old rather than the 11% required by Bank Alpha, whereas good borrowers prefer to pay 5% to the latter rather than 8% to the former. Note that the same adverse selection effect takes place between Bank Alpha and Bank Beta, which has a more granular rating system and, therefore, is able to better differentiate prices. Bad customers borrow from the less sophisticated bank, which attracts risk without being rewarded for it. As soon as risk transforms into losses, this bank is obliged to increase interest rates; in doing so, it even more adversely selects customers and eventually goes bankrupt.

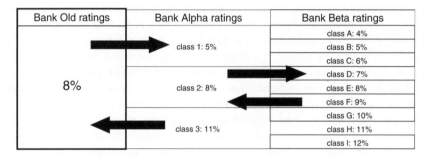

Figure 7.3   Adverse selection.

Of course, the race to increase rating scales granularity has a limit in discriminatory power and effective calibration of rating systems.

In conclusion, not only do we need a rating system fully coordinated with the overall approach to lending, but we also need a rating system powerful enough to compete. As a consequence, for many banks it represents the key innovation to be implemented in the next decade, above all with regards to lending to small and medium enterprises.

7.2   Internal ratings assignment trends

Often, the development of internal rating systems is considered regulation driven. When this is the case, it means that what we have recently underlined in the previous paragraph is not agreed upon or, worse, is not understood. The requirements of Basel II have simply accelerated a trend of improvements and modernization of credit risk management that was already occurring in the more advanced banks. In fact, well before starting to discuss about a future Basel II in 1999, internal rating systems were already born and were spreading in banks; two surveys conducted in the late 1990s confirm this.

The Federal Reserve surveyed large banks in the United States and published results in 1998 (Treacy and Carey, 1998). Two years later, the Basel Committee reported similar findings concerning international banks' rating systems: 'the survey highlighted a considerable number of common elements among rating systems. These include the following: (a) Banks appear to consider similar types of risk factors – such as the borrower's balance sheet, income statement, and cash flow performance – when assigning a rating. However, the relative importance of these factors, and the mix between quantitative and qualitative considerations, differed between the banks surveyed and, in some cases, between different borrower types within the same bank; (b) regardless of whether banks assign ratings to borrowers, facilities, or both, the vast majority of banks surveyed assign these ratings based on assessment of the counterparty. Those banks assigning grades to individual facilities also consider the degree of risk contributed by the specific characteristics of the transaction being rated, while some institutions assigning only borrower grades appear to be considering facility characteristics when allocating economic capital for credit risk; (c) the information gleaned from ratings is utilized (or expected to be utilized) in broadly similar processes at the banks surveyed, including management reporting, pricing, and limit setting. While there does not appear to be a single standard for the structure and operation of internal rating systems, the survey highlighted a few alternative approaches. These can be viewed as points on a continuum with, at one extreme, systems focused on the judgment of expert personnel and, at the other, those based solely on statistical models' (Basel Committee, 2000a, p.4).

At the time, judgment-based systems were much more common than model-based systems. Since then, there has been a clear acceleration in the usage of statistical-based models and the judgmental contribution has been greatly reduced when ratings have been applied to small and medium enterprises.

In 2003, a survey conducted on the eight most relevant Italian Banks (BNL, BPM, Capitalia, Intesa, MPS, Popolare di Lodi, SanPaolo, and Unicredit) showed that: (a) internal rating systems were mainly used in large banks, (b) rating systems date of birth was very recent (Table 7.2), and (c) banks had a clear orientation towards statistical-based models (De Laurentis, Saita, and Sironi, 2004).

The trend towards an increasing mechanization of credit risk analysis technology has developed in parallel with the diffusion of internal rating systems, and has been largely surveyed at the international level since the outbreak of the new century (Brunner, Krahnen, and Weber, 2000; Akhavein, Frame, and White, 2001; Ely and Robinson, 2001; Frame, Srinivasan, and Woosley, 2001; Berger, Frame, and Miller, 2002). More recently, research carried out by the Bank of Italy in 2007 on more than 300 banks (Albareto et al., 2008) illustrated that:

1. rating systems are in operation in almost all large banks, both for SMEs and for larger firms;

2. rating is considered the result of the application of quantitative techniques to credit risk analysis;

3. financial statements, internal behavioral data, and credit register data are the relevant information which drive ratings, especially for SMEs;

Table 7.2   Eight Italian banks internal rating systems: starting date of operation.

Segments	Banks*							
	A	B–H	C	D	E	F	G	Mean
Large corporates	–	2004	2002	2001	1998	2003	2003	2002
SME higher segment	2003	2004	2001	1998	1998	2003	2003	2002
SME lower segment	2003	2004	2001	1998	1998	2003	2001	2002
Small businesses	–	2004	2001	1998	1998	2003	2003	2002
Public sector entities	–	2004	–	2004	2004	2004	2004	2004
Banks	–	2004	2001	2002	2004	2003	2003	2003
Financial institutions	–	2004	2002	2003	2000	2003	2003	2003
Individuals	–	2004	2002	2000	2000	2004	2003	2002
Sovereigns	–	2004	2002	2002	2000	1999	2004	2002

*Banks' names are undisclosed. Letters are randomly assigned to banks.

4. in credit underwriting to SMEs, about 50% of large and medium banks do not leave any possibility to override the output of statistical based models; only 20% of smaller banks take this position;

5. larger underwriting powers have been given to branch managers, thanks to well disciplined ratings processes, which are difficult to manipulate;

6. the internal organization of credit and risk management departments is influenced by banks-specific circumstances and constitutes an essential component of banks' competitive capability.

In conclusion, internal rating systems are critical tools in credit management. In retail segments, characterized by highly fragmented operations and high frequency decisions, it is nowadays rare to do business without statistical-based credit models, largely automatic procedures, and great reliance upon information systems. Also, in the mortgage loans segments, the production technology is now strongly based on mechanical rating assignment approaches. In the SME and large corporate segments, this transformation is still under way, with large banks leading the way; considering large banks' markets shares in almost all countries, we can deduce that model-based techniques in rating assignments already have a great impact on bank-firm relations.

7.3 Statistical-based ratings and regulation: conflicting objectives?

The trend towards an increasing use of mechanical approaches in rating assignment is not driven by the new Basel capital accord. In fact, Basel II is completely neutral on the matter; it requires a much higher degree of formalization of decisions once ratings have been assigned, rather than for rating assignment. Basel II leaves banks free to choose assignment processes, coherently with the overall philosophy of the new capital accord that tends to align regulatory approaches to management approaches, avoiding imposing standard references: 'it is not the Committee's intention to dictate the form or operational detail of banks' risk management policies and practices' (Basel Committee, 2004, p.2). Also Paragraph 417 of the regulation is particularly clear: 'Credit scoring models and other mechanical procedures are permissible as the primary or partial basis of rating assignments, and may play a role in the estimation of loss characteristics'. Therefore, models are neither forbidden nor compulsory. In any case, if banks decide to use models, a number of requirements must be met (listed from Paragraph 417 onwards). The reason is that 'credit scoring models and other mechanical rating procedures generally use only a subset of available information. Although mechanical rating procedures may sometimes avoid some of the idiosyncratic errors made by rating systems in which human judgment plays a large role, mechanical use of limited information also is a source of rating errors... Sufficient human judgment and human oversight is necessary to ensure that all relevant and material information, including that which is outside the scope of the model, is also taken into consideration, and that the model is used appropriately' (Paragraph 417).

This means that rating assignment mechanization is not required by Basel II, but rather it is:

(a) an autonomous choice made by banks, when they are interested in increasing process standardization, cost cutting, delegated underwriting powers (balanced by more objective risk assessments), and personnel orientation to sell (rather than to risk analysis);

(b) a possible choice of national supervisors, when they privilege objectivity of ratings and of aggregate measures of risks over management needs of carefully assessing individual ratings to be applied to each individual borrower and facility.

Focusing on the latter case, we can consider the position of the Bank of Italy that, among other national supervisors, pushes towards a central role of models in rating assignment. Basel II has been re-written in a national regulation (Bank of Italy, 2006) where, initially, all possible rating assignment processes are listed (Title II, Chapter 1, p.62): 'IRB approaches differ in terms of the importance of mechanical outputs generated by the model and those based on the judgment of lending experts. In general, it is possible to distinguish between:

• mechanical systems (which may incorporate standardized qualitative elements), from which discretionary, justified overrides are structurally excluded;

• systems in which mechanical assessments may be modified by experts through the override process with information that cannot easily be standardized or in any event is not considered by the model;

• systems primarily based on the discretionary judgment of an expert.

In selecting the most appropriate organizational approach, banks shall take account of their size, operational characteristics, and organizational structures, as well as the segments of the portfolio concerned (large corporate, corporate, retail), which generally involve different analysis methodologies, procedures and professional roles'.

These statements are fully compliant with Basel II and seem to leave full autonomy to banks whether to assign ratings by more *judgmental* oriented approaches or by more *mechanical* ones. Also, the rating 'replicability requirement' of Paragraph 410 of Basel II is translated into the Italian regulation maintaining a neutral position on rating assignment process orientation (Bank of Italy, 2006, p.63).

However, in other statements of the 2006 regulation, and in particular in the paragraph related to the 'Integrity of rating assignment processes' (p.63), a different view emerges: 'where the bank's IRB system incorporates discretionary assessments by sector experts in assigning a final rating, banks shall adopt appropriate organizational and procedural precautions to ensure the integrity of the rating process, so that the final rating assignment shall not be influenced by the involvement of persons with interests that conflict with the objective of an IRB system to perform an accurate and detailed assessment of the creditworthiness of the counterparty.

Such conflict may emerge where the persons responsible for the final rating assignment: (a) perform an activity that is assessed on the basis of targets for lending volumes or revenues; or (b) have the power to authorize lending decisions.

Where responsibility for the final rating assignment lies with persons involved in the ordinary credit assessment and decision-making process, the organizational arrangements adopted shall ensure that persons with the power to authorize lending decisions or who are affected by the incentive mechanisms described above do not also have any responsibility for final rating assignments.

Appropriate precautions shall also be taken to ensure that the independence of the assessments of the persons who make the final rating assignment is not jeopardized by their belonging to an organizational structure with the power to authorize lending decisions or where remuneration is linked to the achievement of targets for lending volumes or revenues.

Where responsibility for making the final rating assignment is centralized within a specific unit that directly calculates or confirms or modifies ratings generated using a mechanical method or by sector experts, specific attention shall be paid to the following:

- the organizational position of the unit and the extent of its independence from the functions responsible for business growth and lending;

- the position of the unit's manager within the bank hierarchy, the qualitative and quantitative composition of the personnel assigned to the unit and any related incentive mechanisms;

- the rating assignment procedure, in particular the scope of override powers;

- the characteristics of the portfolio, in terms of the number and size of the positions to be assessed by the structure'.

In practice, on one hand, everybody agrees with the need to avoid the final rating decision being made by bank personnel whose objectives conflict with the safeguard of bank safety; on the other hand, it is questionable to classify those having the power to underwrite loans as those having conflicting objectives with bank safety. This statement is not trivial because:

1. It assimilates those having loan underwriting powers with those having commercial objectives, softening the key defensive line of safe bank management that requires the separation between loan officers (belonging to the bank commercial function) and credit officers (belonging to the bank credit function). Even if, in practice, there are great interconnections between the two, commercial functions are typically carried out in loan departments by loan officers or relationship managers who assess customers' needs and sale products, advisory services, and after sale assistance. On the contrary, credit functions include the typical activity of credit officers, such as credit analysis and loan underwriting, that takes place in credit departments and risk management departments. Therefore, the statement that we are debating shifts the defensive line to a third organizational unit, the one which has the power to set ratings' final decisions.

2. It separates the responsibility of rating assignment final decisions from the responsibility to take the risk by underwriting the loan (note that we also refer to loan reviewing processes).

3. It may reduce the informational content of ratings because they may not include all data and evaluations that have led to loan underwriting; in this case, a rating is not a 'ranked synthetic judgment' backing each loan extension decision.

4. It may create a misalignment between the credit officer's opinion and the outcome of the rating process that are difficult to explain to the borrower (who certainly has his own opinion) and to bank personnel (who are more interested with individual cases rather than with the 'average' portfolio performance of the rating system).

In addition, some supervisory authority requires that only small room is left for adjusting the model's results. The so-called 'override' is often only allowed in one direction: downgrading. The term itself expresses exceptionality. In other words, the search for ratings integrity seems to indicate an undisclosed orientation towards a model's based ratings with limited judgmental contributions. This has great implications on the common understanding of what ratings are, both inside banks and for their customers. In fact, it implies that ratings have a regulatory function as well as a portfolio management function, whereas daily lending operations are based on different, more bottoms up, more judgmental, and more traditional creditworthiness evaluation techniques.

However, this hypothesis should be excluded by the ratings 'daily usage' requirement (p.65): 'The rating system is not only an instrument for calculating capital requirements, but it must also be used in conducting the bank's regular business. Accordingly, banks shall be authorized to adopt the IRB approach for calculating capital requirements only if their rating system has an essential role in lending, risk management, internal capital allocation and bank governance functions'.

As a consequence, banks cannot have different rating systems for regulatory and management purposes. Therefore, if there is a disagreement between those making the final rating decisions and those having loan underwriting powers, one of two mutually exclusive situations occur:

1. the final rating disciplines the underwriting decision, shifting a model's result on the client, even if the credit officer disagrees;

2. the underwriting decision is made by taking the final rating only partially into account; in this case, the final rating (and all processes that follow in credit management and regulation) does not fully reflect credit risk evaluations upon which the underwriting decision was taken.

7.4   Statistical-based ratings and customers: needs and fears

The 2008 financial crisis and the subsequent economic crisis have put the role of banks as allocator of financial resources at the very heart of the business debate once again. On one hand, firms have blamed banks for the credit crunch when they needed the most support. On the other hand, banks have been seriously impacted by loan provisions and losses. At the same time, outstanding regulators such as Governor Draghi, chairman of the Financial Stability Board, have invited banks to allocate credit with a forward looking medium term perspective (Draghi, 2009). This is a key request in order to overcome the crisis and create a robust economic growth after the crisis. It directly addresses the issue of banks' credit selection techniques and policies. Of course, a particularly important aspect is the structural modification of bank lending technology concomitant with the adoption of internal rating systems. This is even more important in the SME segment, given that SMEs do not have many alternatives to bank lending.

Lacking sufficient empirical evidence of internal ratings behaviors during the economic cycle and during times of strong discontinuity in the state of the economy, it is useful to use the evidence conveyed by ratings issued by large international rating agencies. In times of difficulty, there is a weaker relation between default rates and rating classes: this is clearly shown by lower Gini ratios. Therefore, a question arises: are currently used internal rating systems able to drive the allocation of bank credit with a long term perspective, facilitate economic growth, and avoid procyclicality? To answer these questions, four points should be discussed:

(1) which phenomena are observable in bank loan markets when internal ratings are introduced; (2) the regulatory guidelines, already examined in the last section; (3) how current internal rating systems are built; and (4) which economic and organizational constraints limit a larger use of *judgmental* approaches, at least as a complement to statistical based models output.

The introduction of internal ratings in the Italian market, which is mainly made by small and medium sized companies, is paradigmatic. Towards the end of the 1990s, an analytical analysis of the SanpaoloIMI Bank loan portfolio was carried out. Customers were segmented according to the size of the consumption of bank services, using the National Credit Register and financial statements for international active borrowers. The market structure that emerged was simple and aligned with expectations:

- One-third of loans were extended to borrowers having SanpaoloIMI as their main bank lender (with a share higher than 75%). For these borrowers the average number of relationships with banks was three (it was rare to find more than five banks) and firm size was quite small.

- One-third of loans were extended to borrowers having SanpaoloIMI as their privileged lender. The bank was covering, on average, 35% of bank borrowing and the number of bank relationships easily reached 10; firm size was medium to high.

- The final one-third of loans was extended to borrowers for which SanpaoloIMI was one of the many non-privileged banks (from 30 to 50 banks). Average firm size was large to very large.

Clearly, SanpaoloIMI was playing different roles: exclusive bank, privileged bank, and product or niche bank.

The same analysis was carried out ten years later, when SanpaoloIMI merged with Intesa. These differences had disappeared. Market share on individual borrowers no longer had any relation with their size or loan amount (R-squared was only 1.2%): significant market shares can be found on larger customers and small market shares on smaller customers.

In the meantime, internal ratings had been introduced in bank operations. The two phenomena can be uncorrelated. But, in any case, ratings have not reduced competitive options! Instead, a larger diversification of competitive choices seems to have taken place. In addition, the use of portfolio models has increased opportunities of actively managing concentration risk, softening limits to lending to single names or sectors.

Further evidence can be obtained by observing loan prices. A widespread belief is that ratings have been used in order to indiscriminately increase interest rates. Once more, empirical data do not support this belief. Consider two analyses carried out before and after the introduction of internal ratings by the SanpaoloIMI Group in 1998. The bank surveyed prices of loans extended in 1997 to firms that were going to be rated the year after, using data from the National Credit Register; the bank was going to rate approximately 15% of non-financial Italian firms and, as the National Credit Register stores data concerning the average prices on loans to borrowers granted by all banks operating in Italy, collected data can be considered representative of average market conditions. The survey illustrated that average prices (credit spreads) were not reflecting firms' risk but rather firms' size and other sales factors.

In 2002, the Bank of Italy started to publish a classification of domestic borrowers by rating class. The 2001 Annual Report (Bank of Italy, 2002) indicated the average credit spreads per rating class reported in Figure 7.4. Using this data, SanpoaloIMI carried out a simulation in order to identify theoretical prices to be applied according to:

- the bank's internal models,

- the bank's internal models, but using the FIRBA framework available at the time for calculating capital absorption.

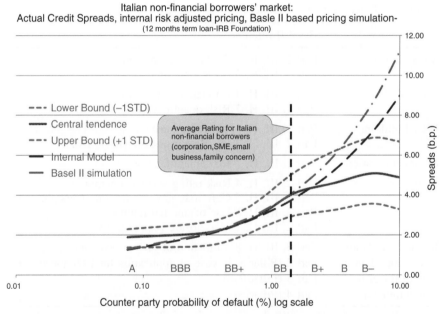

Figure 7.4   Observed and theoretical credit spreads for Italian non-financial firms in 2001. Source: *Maino and Masera (2005)*

Figure 7.4 shows that, unlike the 1997 evidence, the relationship between credit risk and credit prices was strong, even if prices underestimated the risk in worst classes.

Since then, for bank loans, the sensitivity of prices to risk has been refined. At the same time, between 2005 and the first semester of 2007, credit spreads had reached the lowest level since World War II, in a period of abundant credit supply on all maturities and also in the segment of small firms.

Of course, internal ratings and credit risk management are not the trigger factors for all this. But at the same time, it is clear that their introduction has not been historically correlated with higher credit prices, lower credit availability, and shortening maturities. These are only concerns, and do not reflect reality.

7.5   Limits of statistical-based ratings

As indicated in the last three chapters, statistical-based rating systems limits may be derived, first of all, from errors and approximations during model's design, development, and use. These issues are analytically dealt with in the Basel II regulation, above all in Part 2, III, H. A taxonomy of what we can define as 'direct sources of model risk' is reported in Table 7.3, together with references to paragraphs of the regulation where they were examined.

Table 7.3   Direct sources of model risk.

Direct sources	Basel II references
Model design	Part 2, III, H, 3, i) Rating dimension
	Part 2, III, H, 3, ii) Rating structure
Dataset	Part 2, III, H, 4 iv) Data maintenance
	Part 2, III, H, 7 Risk quantification
Model building	Part 2, III, H, 3, iii) Rating criteria
	Part 2, III, H, 3, iv) Rating assignment horizon
Model calibration	Part 2, III, H, 7 Risk quantification
Model use	Part 2, III, H, 4 Risk rating systems operations
	Part 2, III, H, 5 Corporate governance and oversight
	Part 2, III, H, 6 Use of internal ratings
	Part 2, III, H, 12 Disclosure requirements
Internal validation and compliance with Basel II requirements	Part 2, III, H, 8 Validation of internal estimates
	Part 2 (Pillar 1, in case of application for IRB approaches)
	Part 3 (Pillar 2)
	Part 4 (Pillar 3)

Source: De Laurentis Gabbi, 2010

A second source of statistical-based rating systems (SBRSs) limits may derive from a more strategic issue: the decision to use them singularly, with marginal or no contribution of experts' judgments to the assignment of the final rating. SBRSs may be appropriate for many market segments and financial products for which model risks are balanced by the reduction of analysis costs, increased objectivity and comparability of ratings, quicker credit decisions, better separation between loan officers and credit officers. Are SBRSs also appropriate for segments of the credit market traditionally based on relationship banking, such as the SME segment?

The answer is: be very careful (De Laurentis and Gabbi; 2010). In relationship banking, banks are engaged in both assessing borrowers' creditworthiness in the medium to long term and providing customers with the most appropriate products, advisory services, and assistance. For information-based theory of financial intermediation, banks exist because of information synergies they can exploit from credit risk assessment processes and commercial activities. The problem is that risk analysis based on SBRSs does not produce information spillovers which are beneficial for commercial activities.

A different scenario arises when relationship managers and credit analysts interact to elaborate information. There are various degrees of spillovers produced by more or less sophisticated judgmental approaches. A sophisticated approach, in which analysts are required to assign partial ratings to firm's 'risk factors', such as business risks, financial risks, (borrower's own) credit risks, and operating risks, would require integrating all available information sources in order to address the key risks a firm is facing (bottom of Figure 7.5). In doing so, the bank achieves a deep understanding of a firm's strengths/weaknesses, opportunities/needs, and this is also useful for relationship managers who can comprehensively assist and advise customers. At the same time, they can provide valuable private information to credit analysts for a better assessment of risks on a longer time horizon.

A simpler judgmental approach (top of Figure 7.5), where analysts are required to determine the final borrower rating by assigning partial ratings to different

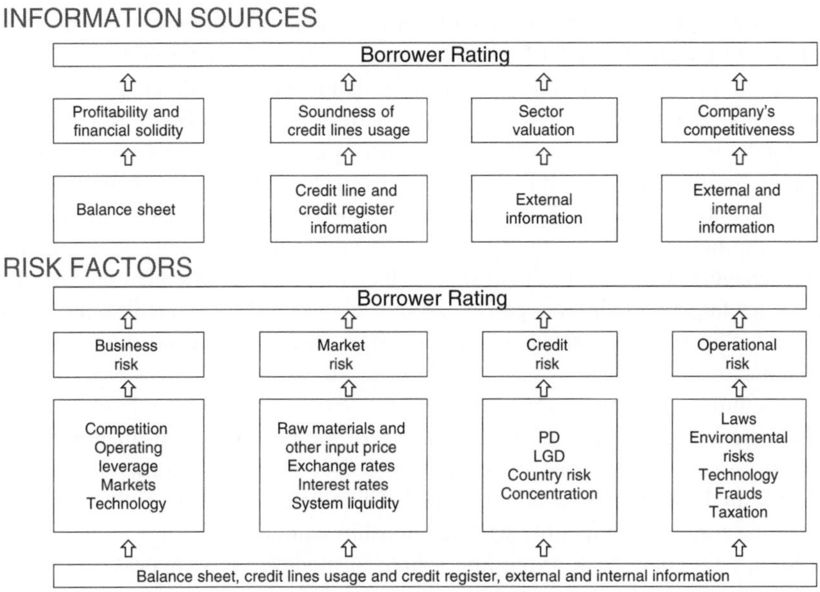

INFORMATION SOURCES

Borrower Rating			
⇑	⇑	⇑	⇑
Profitability and financial solidity	Soundness of credit lines usage	Sector valuation	Company's competitiveness
⇑	⇑	⇑	⇑
Balance sheet	Credit line and credit register information	External information	External and internal information

RISK FACTORS

Borrower Rating			
⇑	⇑	⇑	⇑
Business risk	Market risk	Credit risk	Operational risk
⇑	⇑	⇑	⇑
Competition Operating leverage Markets Technology	Raw materials and other input price Exchange rates Interest rates System liquidity	PD LGD Country risk Concentration	Laws Environmental risks Technology Frauds Taxation
⇑	⇑	⇑	⇑
Balance sheet, credit lines usage and credit register, external and internal information			

Figure 7.5   Alternative judgment based approaches.

'data sources' (typically, income statement, balance sheet, flow of funds statement, behavioral data, credit register data, business sector, strategic positioning) would result in a poorer understanding of firms, because the analysis misses the key risk factors which trigger credit risk: information spillovers for commercial activities are much lower and credit risk assessment is less forward looking. However, when compared with the outlined simpler judgmental approach, SBRSs are even less informative.

The hypothesis of banks requiring the use of judgmental analysis of borrower creditworthiness for purposes other than rating assignment and credit underwriting suffers from severe limitations: (1) bearable costs (and consequently effectiveness) of this analysis because of its limited scope, (2) low incentives to undertake an in-depth analysis of a firm, and (3) relationship managers do not benefit from credit analysts' expertise and bear the entire burden of the analysis. This is the framework in banks using SBRSs as the key tool to assess probabilities of default.

In case the judgmental analysis is developed for credit underwriting, whereas SBRS are used for other risk management purposes, the drawbacks are: (1) ratings do not benefit from judgmental analysis achievements, as they are separate processes; and (2) ratings do not reflect risk valuations developed for underwriting decisions; that is, provisions, capital requirements, and risk-adjusted performance measures based on rating are not directly linked with individual lending decisions.

In conclusion, SBRSs do not create information synergies that are crucial for relationship lending. Instead, judgment-based approaches to creditworthiness evaluations, and above all the most advanced ones based on the identification of firms' risk factors, are able to produce information that is useful for commercial activities and borrower risk control at the same time.

To overcome some weaknesses of SBRSs, banks are often combining model results with human judgment ('override process'). It seems appropriate to also take into account all relevant and material information not considered by the model, which is a specific Basel II requirement. However, in the Basel II framework, for 'model-based ratings', overrides are strictly regulated exceptions. Banks are actually limiting the possibility to make overrides, above all when it leads to improvements of ratings. The rationale behind limiting changes to model-based ratings through overrides is twofold: override proposals are usually set by relationship managers, who can be interested in increasing loans extended and reaching their sales targets; qualitative considerations, considered for overrides, are simplified valuation of some aspects of the borrower, usually based on multiple-choice questionnaires.

In summary, if SBRSs are completed by overrides that are not based on a true and comprehensive judgment-based analysis of the borrower; information production on borrowers risks and opportunities, useful both for commercial purposes and risk analysis purposes, is immaterial. Therefore, we conclude that also these approaches are not compatible with relationship lending.

Now, let's come back to the first source of model risk: weaknesses embedded into the model itself, hampering its ability to rank risk (discrimination), and to produce a strong and stable relationship with default rates (calibration).

Natural benchmarks for internal ratings are ratings issued by international agencies, because of their wide accessibility and long time series. Table 7.4 shows that, in the long run (1981–2008), Standard & Poor's ratings meet the three conditions usually required for rating systems: default rates are increasing as rating class deteriorates, the frequency of borrowers maintaining the same class is high, rates of migration to closer classes are higher than rates of migration in farer classes (Standard & Poor's, 2009a)[1]. Of course, increasing the time horizon, decreases rating stability, but the two other conditions are still met (Table 7.5).

Table 7.4   Corporate Transition Matrices (1981–2008), One-Year Transition Rates (%).

Da/a	AAA	AA	A	BBB	BB	B	CCC/C	D	NR
AAA	**88.39**	7.63	0.53	0.06	0.08	0.03	0.06	0.00	3.23
AA	0.58	**87.02**	7.79	0.54	0.06	0.09	0.03	0.03	3.86
A	0.04	2.04	**87.19**	5.35	0.40	0.16	0.03	0.08	4.72
BBB	0.01	0.15	3.87	**84.28**	4.00	0.69	0.16	0.24	6.60
BB	0.02	0.05	0.19	5.30	**75.74**	7.22	0.80	0.99	9.68
B	0.00	0.05	0.15	0.26	5.68	**73.02**	4.34	4.51	12.00
CCC/C	0.00	0.00	0.23	0.34	0.97	11.84	**46.96**	25.67	14.00

Source: Standard & Poor's (2009a).

Table 7.5   Corporate Transition Matrices (1981–2008), Five-Year Transition Rates (%).

Da/a	AAA	AA	A	BBB	BB	B	CCC/C	D	NR
AAA	**54.23**	23.49	5.10	0.93	0.12	0.09	0.06	0.28	15.69
AA	1.75	**51.73**	23.52	4.08	0.60	0.36	0.04	0.30	17.62
A	0.12	5.92	**53.37**	15.23	2.44	0.95	0.17	0.68	21.11
BBB	0.05	0.78	10.84	**47.07**	8.28	2.91	0.52	2.57	26.99
BB	0.02	0.12	1.51	12.26	**28.12**	11.03	1.59	9.98	35.37
B	0.03	0.06	0.50	2.13	10.92	**20.83**	2.88	23.18	39.47
CCC/C	0.00	0.00	0.23	1.21	3.48	11.21	**3.33**	47.80	32.73

Source: Standard & Poor's (2009a).

Discriminatory power (in terms of Gini ratios) also decreases when a longer time horizon is considered, as it is more difficult to forecast issuers' quality after five years from rating assignment rather than after only one year (Figures 7.6 and 7.7). However, Standard & Poor's ratings present, in any case, satisfactory performance for corporate borrowers: Gini coefficients are fair both at one year and at longer time horizons, above all, for non-financial firms (Table 7.6)

Now, how do banks' rating systems and, in particular, SBRSs perform compared with agencies' ratings? Gini coefficients are sample-dependent measures. Moody's has carried out one of the few studies that benchmarks SBRSs on a

---

[1] Moody's transition matrices are reported in Section 3.1.4.

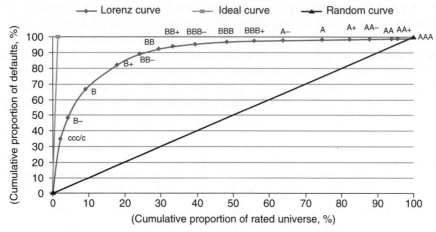

Figure 7.6   Global One-Year Relative S&P's Corporate Ratings Performance (1981–2008). Source: *Standard & Poor's (2009a).*

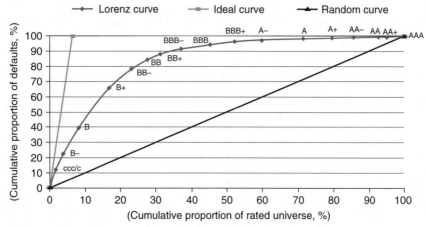

Figure 7.7   Global Five-Year Relative S&P's Corporate Ratings Performance (1981–2008). Source: *Standard & Poor's (2009a).*

common dataset (Moody's Investors Services, 2000): most of the tested models had a one-year Gini ratio in the range of 50 to 75% for out-of-sample and out-of-time tests. The solution, envisaged by supervisory authorities of creating a reference dataset of borrowers to be used to benchmark rating systems performances is perturbed by the different structures of independent variables, in particular, internal behavioral and qualitative data.

Qualitative data create the biggest problem for comparisons. If they are considered on a merely judgmental basis, it is not possible to replicate the rating on a large scale: this is the case for overrides of SBRS based on experts' judgments. In SBRSs qualitative data can be incorporated as nominal and ordinal variables,

Table 7.6   Global Corporate Gini ratios (1981–2008).

Sector	– Time horizon (years) –			
	1	3	5	7
Financial				
Weighted average	77.76	69.92	64.12	61.88
Average	78.53	72.43	66.08	61.81
Standard deviation	(23.94)	(14.09)	(13.80)	(11.00)
Nonfinancial				
Weighted average	84.32	78.87	76.24	74.47
Average	82.95	76.64	73.20	70.22
Standard deviation	(6.51)	(5.11)	(5.16)	(4.98)

Source: Standard & Poor's (2009a).

usually collected by closed-form questionnaires filled in by relationship managers. Consistency issues in the treatment of qualitative information is greatly reduced when relationship managers are just required to fill in a questionnaire, but it still remains for questions with subjective answers and in the presence of incentives schemes based on the amount of loan 'sold' (as internal analyses conducted by many banks have shown).

Qualitative data collected by questionnaires may participate to the final model in two different ways (De Lerma, Gabbi, and Matthias, 2007). The first approach is to use individual nominal and ordinal data as possible explanatory variables, together with quantitative data, in the estimation of the final algorithm representing the SBRS. This approach is rare because: (1) most qualitative data is either nominal or ordinal, so they are crowded out by variables selection procedures because of their low discriminatory power; (2) the collection of questionnaire-based qualitative data has been implemented only recently, therefore datasets are small; and (3) different types of quantitative data are not always available for all borrowers. This is why banks tend to use a second approach: a specific 'qualitative module' is built; it produces a 'qualitative rating' to be subsequently combined with 'partial ratings' obtained by other 'quantitative' modules (typically for financial statement, credit register, and internal behavioral data). Using this approach, more qualitative data is indirectly entered into the SBRS. But the nature of qualitative data leads to a low Gini ratio of the qualitative module. Thus, when combining partial ratings into the final algorithm, its relative relevance is low. The final result is that SBRSs do not leverage much on qualitative information and, therefore, do not give enough weight to strategic and competitive profiles that, on the contrary, can be crucial in times of great discontinuities of the economic cycle.

Quantitative data entered into SBRSs may either derive from the past or represent economic forecasts. The use of a forecast as explanatory variables in SBRSs appears to be rare and, if present, it is more common to include them in the qualitative module rather than to mix them up with historical, more objective, and certified data.

To evaluate ratings performance, a key aspect is to clarify the time frame by which quantitative objective information is collected. Different sources of quantitative objective information have a different newness. Is it positively correlated with predictive power? The answer is yes, if we consider a one-year observation period. In fact, in this case, the typical result when calculating ROC curves for the accounting module, credit register module, and behavioral module, and then for the final model that combines them (using partial scorings as explanatory variables), is depicted in Figure 7.8. The behavioral module has the best performance among all partial modules. Performances are only slightly improved when it is combined with other modules in the final model. A final model that closely resembles results obtained from the behavioral module is a great source of model risk, in terms of short sightedness.

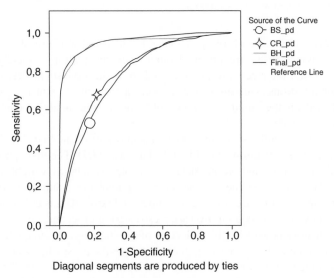

Figure 7.8   ROC curves of different modules of a SBRS and of the model as a whole.

In fact, behavioral data suffer from two severe limitations: (1) they only reflect the current conditions of debts for the borrower (in other words, they reflect what is going on at the specific point in time, from the crude perspective of debt balances and cash flows, ignoring value creation and firm's potential); and (2) they are reflected information because they depend on a bank's own credit decisions: if a bank extends more credit to a given borrower, the debt behavior of the borrower improves. At a lower level, credit register data also suffer for the same weaknesses. In the end, the final model relies on point-in-time, short-term, and reflected information.

Note that, when the observation period is one year, behavioral data (and, to a lesser extent, credit register data) show a higher discriminatory power than financial statement data. This is why simultaneously considering all types of quantitative data in the long list used to estimate a model leads to an almost complete crowding out of financial statement data by behavioral and credit register data. On the other hand, when different modules are developed for different types of data, the typical outcome is described in Figure 7.8. Therefore, this approach increases, more in theory than in practice, the variety of information on which the model relies. In fact, the final model resembles self-determined and very point-in-time information.

The majority of banks use a one-year observation period from time zero as the time frame for building SBRSs to be used in daily operations, even if this is not Basel II compliant (§414 of the regulation states that 'although the time horizon used in PD estimation is one year, banks must use a longer time horizon in assigning ratings'). If a two-year or three-year period of observation is considered, behavioral and credit register data lose their apparent discriminatory power, and accounting data gain relevance in the final SBRS.

From this analysis, a further question arises. Do SBRSs perform better or worse than judgment-based ratings in periods of economic turnarounds? No robust analyses exist, because SBRSs are very recent, dating back to only the late 1990s for a few banks and much later for other banks; and SBRSs are continuously improved, often so much that it is impossible to back test them on old datasets. However, considering SBRS's structural characteristics, one can conclude that they are more point-in-time and are based on a much smaller set of variables; hence, their performance probably lags behind judgment-based approaches in rapidly changing times.

In conclusion, SBRSs are a great source of model risk, in particular in cases of relationship lending and economic cycle shocks. When credit decisions concerning individual borrowers depend on SBRSs, their discriminatory capability and their calibration are required to hold at a satisfactory level, case by case, and on longer (than one-year) time horizons. Whereas, for transaction-based lending and for other applications, such as calculating bank provisions and capital adequacy, good Gini ratios on vast aggregates obtained by point-in-time rating systems can be sufficient.

7.6 Statistical-based ratings and the theory of financial intermediation

The theory of relationship banking points out that, ceteris paribus, strong bank–firm relationships produce a number of positive effects for players and the economic system as a whole:

- higher credit availability, above all in the phases of firms' structural breaks (large investments, strategic turnaround, mergers and acquisitions, accelerated internal growth, etc.),

- lower requirements of collaterals and guarantees,

- lower cost of credit,

- increased borrowers' transparency and lenders' forward looking perspective in credit risk analysis,

- more stable loans prices and credit availability during the economic cycle,

- better advising and consulting services offered by banks on firms financial and non-financial decisions.

In mainstream literature, a key factor for relationship banking is a credit risk assessment process based on the use of soft information, bottom up methodologies, and customer proximity of those having lending authority (Diamond, 1984; Berger and Udell, 2001; Petersen and Rajan, 2002). Berger and Udell (2001) consider 'four main distinct lending technologies' – financial statement analysis, asset-based lending, credit scoring, and relationship lending. The first three lending technologies are often referred to as transactions-based lending, under which the lending decisions are based on hard information that is relatively easily available. Brunner, Krahnen, and Weber (2000) have directly addressed the following questions: is information production changing in credit relationships due to the increasing use of internal rating systems? Is the corporate banker involved in credit rating issuing, reviewing, and control and what is the nature of the rating assignment process? This is a key issue if 'the technology of relationship lending is based on the accumulation of information over time through contact with the firm, its owner, and its local community on a variety of dimensions. The information is often "soft" data that may be difficult to quantify verify and transmit through the layers of management' (Berger and Udell, 2001).

We consider SBRSs that are insufficiently integrated in the credit analysis process that takes place in credit departments, and whose outputs are insufficiently completed by human judgment, capable of significantly transforming the way banks compete in SME segments. In front of important advantages in terms of operating costs, objectivity, integrity, replicability, and compliance with national regulations, rating systems that are based almost exclusively on statistical based models:

1. reduce the capability of banks to evaluate the specific conditions of a borrower, taking account of its business mix, its development stage, its projects, its competitive position, the outlook of business sectors in which it operates, the current and expected state of the economy (De Laurentis, 2001, p.216 onwards);

2. narrow the role of loan and credit officers, un-incentivize the development of financial analysis competences. Year after year, bank personnel are destined to lose the capacity to read as an integrated framework different information sources and to appraise the sustainability of strategic and operating choices borrowers are undertaking. As evaluating credit risk on a judgment basis, understanding financial needs and advising firms are all based on common information, analytical approaches and expertise, then, the degree of customization of bank services is destined to decrease, whereas their commoditization is destined to increase. Instead, the search for know-how spillovers (intra-customer, intra-product and intra-geography) seems to be a key driver of banks organizations (Baron and Besanko, 2001);

3. diminish the capacity of acquiring soft information from relationship lending; bank-specificity within the suppliers of financial resources fades; commoditization of credit extends to small and medium enterprises, contradicting the theoretical models of relationship banking (Rajan, 1992; Petersen and Rajan, 1994; Boot, 2000; Boot and Thakor, 2000);

4. modify credit availability and cost. Actually, available empirical analyses do not always confirm mainstream expectations. Frame, Srinivasan, and Woolsey (2001) surveyed the behavior of 99 of the most important 200 American banks in lending, pointing out that banks using scoring systems have a larger part of their loan portfolios invested in small businesses (firms having less than 500 employees). According to Berger, Frame, and Miller (2002), 'rules banks', which more directly link loan decisions with scorings (often bought from external vendors) for small business credits (less than US$ 100 000), had increased the volume of credit granted in recent years, even if at higher prices and with higher risks; on the contrary, 'discretion banks' using scoring systems (mainly internally developed) to complement traditional judgment-based approaches to credit risk analysis, had not increased the volume of credit extended, had reduced the cost of risk, and increased prices more than rules banks, probably compensating higher cost of analysis;

5. reduce banks' ability to smooth out the variability of economic cycles, transmitting the changes in borrowers' probability of default more directly to the economic system. In fact, up to now, poorly structured credit analysis processes have allowed the remuneration for credit losses suffered in worst periods to be smoothed out over time (whereas market-based financial systems have always immediately transmitted credit losses to the economy, but distributing them to a much larger number of investors). With the adoption of mechanical approaches to credit analysis and the increased sensitivity of capital absorption (both economic and regulatory capital), as well as with the new accounting standards more orientated to fair values, bank lending assumes a much more pro-cyclical character.

The five points listed above, together with other phenomena, such as the increase in banks size and foreign banks presence (Berger, Klapper, and Udell, 2001), impoverish the bank–firm relationships, with possible negative effects on the capability of banking systems to optimize credit allocation and to help long term economic growth. Vice versa, credit risk analysis approaches, which have a more judgmental orientation, allow credit analysts to evaluate the relevance of different risk factors case by case, consent more customized bottom up analyses, produce bi-directional information spillovers between sales activity and credit risk assessments, require and incentivize higher competences of bank personnel and higher sophistication of compensation schemes.

A question arises: does a bi-directional relationship between rating systems choices (from purely statistical-based to purely judgmental), bank size, and bank organizational structure exist?

Economic literature is abundant but controversial.

On one hand, some large banks, which are traditionally less inclined to do business with small and opaque firms with a relationship-orientated approach, have seen divisionalization by customer segments as the way to attack the attractive markets of local banks (De Young, Hunter, and Udell, 2003). In fact, these are attractive but have become highly exposed to the competitive choices of large banks: 'a community bank business model that emphasizes personalized service and relationships based on soft information is likely to be viable in the long run . . . the survival of community banks in the future depends on the ability of large banks to increase personalization and customization of their services, while still maintaining their low unit cost advantage' (DeYoung, Hunter, Udell, 2003).

On the other hand, in research carried out in 2002–2003 (De Laurentis, 2005) in which 12 European banks were analyzed, they were evenly divided between those affirming that divisionalization by customer segments was undertaken having a vision of corporate banking as customization-oriented business, and those declaring that the reorganization was backed by a vision of corporate banking as a commodity-orientated business. No correlation was observable between the corporate banking vision on one side and credit risk evaluation approaches on the other side. In particular, no empirical confirmation was found for the assumption that: (a) the customized-relationship approach is associated with a growing exploitation of credit risk analyses primarily based on analysts' skills, comprehensive quantitative and qualitative analysis of customers accounts and strategies, medium to long term horizon forecasts; (b) commoditization-orientated approach is associated with analyses aimed at shorter term objectives and substantially relying on mechanical systems. On the contrary, the customization-orientated supply model seemed to be implemented exclusively in order to reconfigure commercial processes (by increasing the commercial specialization of relationship managers); on the credit function side, the mechanization of rating assignment processes was implemented in order to balance the increased customer proximity and commercial orientation of relationship managers.

In other words, while Udell (1989) empirically verified that banks who delegated greater lending authority to loan officers were actually investing more on their control through the loan review function, in the new context characterized by organizational divisionalization and by the introduction of rating procedures, banks seem to exploit the standardization of the loan approval process as a tool for simplifying credit supply. These results have also recently been empirically observed in Italy (Albareto *et al.*, 2008).

These and other pieces of literature do not give a definitive answer to the question. Probably, the way banks are implementing their strategy, their organization, and their analytical tools requires a much larger alignment before final conclusions can be drawn on the value of different business models. For the time being, we can note that, on one hand, customers do not recognize a meaningful added value brought by relationship managers specialized in corporate customers and, on the other hand, the wide range and the frequently renewed formulation of strategies and organization models of corporate banking by almost all banks highlight the low satisfaction that they themselves feel. Therefore, it is not surprising that the 2008 financial crisis has strained bank–firm relationships and both sides are strongly debating around the nature of rating systems and the way they are incorporated into credit decision processes.

The relationship between SBRSs and the theory of financial intermediation is becoming even more complex because new rating agencies for small and medium enterprises are entering into the market. They are offering unsolicited ratings based on SBRSs, being recognized by national supervisors as External Credit Assessment Institutions (ECAI) and, therefore, are contributing to define regulatory capital requirements for banks that have chosen the standardized approach within the three options of Pillar I of Basel II, and, eventually, determining a demand for unsolicited statistical-based ratings by banks and borrowers themselves.

It is worth noting that the business model of rating agencies is well defined and robust when large ticket debt operations are carried out (mainly issuing bonds) by large corporations. These companies know that they can benefit from lower interest rates by choosing these alternatives, rather than the plain vanilla bank borrowing. As the size of borrowing is huge, a small benefit in interest rates produces a large saving. Therefore, large corporations are in the position to pay high fees to rating agencies to assure investors of their creditworthiness; in turn, these fees allow agencies to bear the relevant costs of analysis and drilling into borrowers current and perspective financial and strategic conditions. The need to save the reputation counterbalances the possible conflict of interest arising from borrowers who pay the raters. Reputation is also a barrier to entry into the rating business that it is critical to grant to rating agencies enough market power to utilize with borrowers in order to have wide and timely access to confidential information. Furthermore, reputation pushes agencies to consolidate their relationships with investors:

1. maintaining constant rating architecture and meaning over time,

2. disclosing rating performance,

3. clarifying rating processes,

4. motivating rating decisions.

This business model does not work for rating SMEs. The small scale of operations, the lack of market alternatives to bank borrowing, and the trivial interest savings do not allow investing enough in information acquisition and analysis; as a consequence, only SBRSs can be conveniently used by ECAIs. Now, banks tend to have more information on their SME customers than agencies for many reasons: customer proximity, the offer of a wide variety of services, behavioral data, and credit register data. Therefore, these statistical-based external ratings are a form of outsourcing of credit analysis that is allowed for regulatory purposes, but that should not be erroneously extended to management purposes. In the latter case, it reproduces the same problems previously pointed out in general for SBRSs, also for small banks and their SME customers. The possible negative effects in terms of bank–firms relationships are, in this case, even larger because of the small and local nature of both banks and firms involved. Not to mention that the free riding behavior may increase competition among banks, may lower their economic possibility to collect and evaluate information, and in the long run may erode the basis for 'relationship banking' of small banks.

In the document 'A new capital adequacy framework,' (Basel Committee on Banking Supervision, June 1999) which started the process that lead to the Basel II regulation, the Basel Committee was perfectly aware of all this; in fact, in Paragraph 43 it states that: 'The Committee recognises that internal ratings may incorporate supplementary customer information which is usually out of the reach of an external credit assessment institution, such as detailed monitoring of the customers' accounts and greater knowledge of any guarantees or collateral. Thus, in offering a parallel alternative to the standardised approach based on internal ratings, the Committee hopes that banks will be encouraged to further develop and enhance internal credit risk management and measurement techniques, rather than place an unduly broad reliance on credit assessments conducted by external credit assessment institutions'.

This position was coherently transferred into stringent requirements for eligible ECAIs in the Basel II regulation, at Paragraph 91. Instead, the European Directive and national regulations have left the door open for rating agencies of SMEs which produce unsolicited ratings by SBRSs. Political reasons may have overtaken economic considerations.

7.7   Statistical-based ratings usage: guidelines

Ratings are increasingly diversifying their uses, nowadays ranging from loans origination and underwriting to risk control, underwriting powers delegation schemes, regulatory and economic capital absorption, loans impairment and provisioning, capital allocation and strategic management of banks. All this is in coherence with the broader development of risk management as a driver for creating shareholders value in banks (Resti and Sironi, 2007). But, as uses diversify, ratings have increasing difficulty in being simultaneously adequate for these misaligned and sometimes conflicting purposes (Table 7.7). Therefore, the wide range of uses creates negative repercussions for ratings functionality to adequately reach such diversified objectives.

There are three areas of intervention. The first area concerns the technicality of model development: it is necessary to avoid committing too many errors and approximations in the many methodological traps of model development processes. Moreover, it is necessary to avoid producing a black box whose internal technical logics and economic meanings are unknown or, even worse, contrasting widespread expectations and long-standing expertise. In addition, the bank has to follow a process of cultural advancement, internalizing key competencies of model development, validation, control, and improvement. Eventually, some critical errors in model design should be avoided, such as confusing the time horizon of rating

Table 7.7   Ratings uses and desired characteristics.

Rating uses	Useful characteristics of ratings		
	1. Counterparty-risk discrimination	2. Sensitivity to economic cycle	3. Rating stability and far-sightedness
a. Underwriting (loan by loan)	Yes	High	Low
b. Customer relation	Yes	Low	High
c. Commercial policy	Yes	Average	Average
d. Early warning / watch list	Yes	Very high	Null
e. Risk control and reporting	Yes	High	Average
f. Provisioning (current IAS)	Yes, for general provisions	High	Null
g. Provisioning (IAS after G20 recommendations)	Yes, for general provisions and for yearly EL calculation	Average	Average
h. Economic capital	Yes	High	Low
i. Regulatory capital (*Pillar 1*)	Yes	Average	Average
j. Capital adequacy (*Pillar 2*)	Yes	Low	High

Legend:

1. Counterparty-risk discrimination:	Correctly ranked default rates per class (statically and over time), good granularity of rating scale
2. Sensitivity to economic cycle:	Point in Time ratings, that is to say, depending on the stage of economic and credit cycles
3. Rating stability and far-sightedness:	Through the Cycle ratings, that is to say, independent from the stage of economic and credit cycles, depending only on long term idiosyncratic fundamentals of the borrower
a) Underwriting (loan by loan)	Loan by loan evaluation of credit risk, also functional to loan pricing
b) Customer relationship	Customer relationship management, focused on medium term horizon and on a win-win value creation strategy
c) Commercial policy	Portfolio credit analyses, with profitability objectives typically targeting a 1-year time horizon

(continued)

Table 7.7   *(continued)*

d) Early warning / watch list	Daily monitoring of a borrower's conditions aimed at selecting a watch list of cases to focus on and possibly triggering risk-mitigation actions
e) Risk control and reporting	Portfolio views of credit risk
f) Provisioning (current IAS)	Calculation of expected but not materialized losses for 'loans and receivables' and 'hold-to-maturity' portfolios. Portfolio fair value estimations for financial statements integrative reports
g) Provisioning (IAS after G20 recommendations)	Expected loss approach for portfolios subject to amortized cost approach. Portfolio fair value estimations for financial statements integrative reports. Impairments (IASB, 2009).
h) Economic capital	VAR-type calculations for portfolio and its segments for a proactive management of credit risk
i) Regulatory capital (Pillar 1)	Basel II, Pillar 1 regulatory compliant aggregate calculations for credit portfolio and its segments
j) Capital Adequacy (Pillar 2)	Basel II, Pillar 2 regulatory compliant aggregate calculations for credit portfolio and its segments

assignment with that of rating quantification and early warning systems for daily credit monitoring with SBRSs for loan underwriting and reviewing.

The second area of intervention concerns the organizational profiles of rating systems. They should be aligned with the overall vision-mission-strategy of the bank in the segments of loan markets in which it competes. In fact, ratings are a key component of lending strategies as well as of the broader bank-firm strategies. If a bank is orientated towards relationship banking in a given market segment, it should avoid renouncing to ratings far-sightedness and stability; it should incentivize the collection and evaluation of soft information to be used in both commercial and advising activities as well.

The third area is in the hands of supervisory authorities which should address questions regarding the uniqueness of ratings to be used for management and regulatory purposes, and the suitability of rating assignment processes for both purposes. The hypothesis to have different rating systems for different uses (with a stringent definition of applications' domains) is less heretical than it could appear at first glance, and it goes along with recent similar conclusions on market risks and VAR measures adequacy (Finger, 2009b).

The first area has extensively been analyzed throughout this book and presents a number of dilemmas, not only between management applications and regulatory applications, but also within the regulatory side. For instance, recall the Basel II requirements for ratings time horizons (from Paragraph 414 onwards): 'Although the time horizon used in PD estimation is one year... banks are expected to use a longer time horizon in assigning ratings... A borrower rating must represent the bank's assessment of the borrower's ability and willingness to contractually perform despite adverse economic conditions or the occurrence of unexpected events... The range of economic conditions that are considered when making assessments must be consistent with current conditions and those that are likely to occur over a business cycle within the respective industry/geographic region'. Now compare it with the new IAS39 provisioning approach using the work of Burroni *et al.* (2009): 'while the expected loss model is based on concepts that recall the Basel II framework, they do not match the IRB measures, which are point-in-time (not long-term averages) and forward-looking. In the IRB approach, the expected losses are based on current PDs and (downturn) LGDs, while the expected losses used for dynamic provisioning are long-term averages of losses recorded in the past. The two definitions tend to be closer when banks adopt through-the-cycle rating systems, which is not necessarily the case. It is also worth highlighting that the IRB definition of expected loss is the one to be used for determining the eligibility of general provisions in supervisory capital'.

The second and third areas are interconnected. For regulatory purposes, it is sufficient that ratings provide fair risk forecasts on large aggregates. Hence, regulators are primarily interested in objective mechanical rating assignment processes based on SBRSs and limited, strictly regulated, overrides. On the contrary, from management's point of view (not to mention from the single borrower's point of view) it is necessary that ratings are appropriate for any single case as much as possible and, above all, if advanced applications such as risk-adjusted pricing schemes are in place. In particular, for relationship-orientated banks, the huge investment in privileged information and relationship protocols determines a natural demand for including all available information in the borrower rating. In fact, the latter also is a relevant topic in the dialogue with the customer.

The Basel II position is fascinating from a theoretical point of view: capital adequacy regulation includes and adopts the very same tools that the bank is using for managing credit risk and the lending business. However, if this signifies in practice

- conditioning banks freedom on choosing rating assignment processes optimally aligned with its vision-mission-strategy on lending (by pushing to limit room for judgmental analysis and/or to separate credit department from the organizational unit that makes the final decisions on ratings),

- limiting the relationship-based model of bank–firm relationships and the capacity of understanding borrowers' problems and potentialities (by minimizing, sometimes annihilating, the room for including soft, privileged, forward-looking information in ratings produced by mechanical procedures),

then, capital regulation should be redefined by rearranging the role of the three pillars. There are two alternative ways.

The first way: Pillar 1 IRB approaches are fed by rating systems and default risk measures that are completely centered on credit management needs and uses. In this case, the spirit of the new capital accord is retained but capital requirements calculated according to Pillar 1 are simply a reference point; supervisors would carefully verify, according to the second Pillar and using all available models' back testing and calibration techniques, the adequacy of banks' capital and require a 'bank-specific correction'.

The second way: the objectivity of rating assignment processes is privileged for Pillar 1, adopting SBRSs with limited or no room for overriding; on the other side, a separation between 'regulatory ratings' and 'management ratings' is allowed, sacrificing the spirit of the new capital accord. 'Management ratings' have a higher judgmental content, the 'final rating' is set by the same units or personnel having loan underwriting powers; full convergence between measures of risk for a specified borrower and individual lending decisions is assured. In this case, Pillar 2 has fewer responsibilities for capital adequacy; it is mainly involved in checking the adequacy of rating systems. This solution renounces the fascinating convergence between management and regulatory rating systems, but recognizes that this separation is *de facto* realized when the final decision on rating assignment is not in the hands of those underwriting loans.

In conclusion, there are still a number of technical, organizational, and strategic open issues in building, validating, and using internal ratings. Nevertheless, no bank can adequately compete without using internal ratings.

Bibliography

Arbib, M. A. (Ed.) (1995), The Handbook of Brain Theory and Neural Networks, The MIT Press.

Adelson, M., and Goldberg, M. (2009), On The Use Of Models By Standard & Poor's Ratings Services, www.standardandpoors.com (accessed February 2010).

Akhavein, J., Frame, W. S., and White, L. J. (2001), The Diffusion of Financial Innovations: An Examination of the Adoption of Small Business Credit Scoring by Large Banking Organization, The Wharton Financial Institution Center, Philadelphia, USA.

Albareto, G., Benvenuti, M., Moretti, S. *et al.* (2008), L'organizzazione dell'attività creditizia e l'utilizzo di tecniche di scoring nel sistema bancario italiano: risultati di un'indagine campionaria, Banca d'Italia, *Questioni e Economia e Finanza,* **12.**

Altman, E. I. (1968), Financial Ratios, Discriminant Analysis and Prediction of Corporate Bankruptcy, *Journal of Finance,* **23** (4).

Altman, E. I. (1989), Measuring Corporate Bond Mortality and Performance, *Journal of Finance,* **XLIV** (4).

Altman, E. I., and Saunders, A. (1998) Credit risk measurement: Developments over the last 20 years, *Journal of Banking and Finance,* **21.**

Altman, E., Haldeman, R., and Narayanan P. (1977), Zeta Analysis: a New Model to Identify Bankruptcy Risk of Corporation, *Journal of Banking and Finance,* **1.**

Altman, E. I., Resti, A., and Sironi A. (2005), Recovery Risk, Riskbooks

Bank of Italy (2002), Annual Report 2001, Rome.

Bank of Italy (2006), New Regulations for the Prudential Supervision of Banks, Circular 263, www.bancaditalia.it (accessed February 2010).

Baron, D., and Besanko, D. (2001), Strategy, Organization and Incentives: Global Corporate Banking at Citibank, *Industrial and Corporate Change,* **10** (1).

Basel Committee on Banking Supervision (1999a), Credit Risk Modelling: Current Practices and Applications, Basel, Switzerland.

Basel Committee on Banking Supervision (1999b), Principles for the Management of Credit Risk, Basel, Switzerland.

Basel Committee on Banking Supervision (2000a), Range of Practice in Banks' Internal Ratings Systems, Discussion paper, Basel, Switzerland.

Basel Committee on Banking Supervision (2000b), Credit Ratings and Complementary Sources of Credit Quality Information, Working Papers 3, Basel, Switzerland.

Basel Committee on Banking Supervision (2004 and 2006), International Convergence of Capital Measurement and Capital Standards. A Revised Framework, Basel, Switzerland.

Basel Committee on Banking Supervision (2005a), Studies on Validation of Internal Rating Systems, Working Papers 14, Basel, Switzerland.

Basel Committee on Banking Supervision (2005b), Validation of Low-default Portfolios in the Basel II Framework, Newsletter 6, Basel, Switzerland.

Basel Committee on Banking Supervision (2006), The IRB Use Test: Background and Implementation, Newsletter 9, Basel, Switzerland.

Basel Committee on Banking Supervision (2008), Range of Practices and Issues in Economic Capital Modeling, Consultative Document, Basel, Switzerland.

Basel Committee on Banking Supervision (2009), Strengthening the Resilience of the Banking Sector, Consultative Document, Basel, Switzerland.

Basilevsky, A. T. (1994), Statistical Factor Analysis and Related Methods: Theory and Applications, John Wiley & Sons Ltd.

Beaver, W. (1966), Financial Ratios as Predictor of Failure, *Journal of Accounting Research,* **4**.

Berger, A. N., and Udell, L. F. (2001), Small Business Credit Availability and Relationship Lending: the Importance of Bank Organizational Structure, US Federal Reserve System Working Papers, Washington, DC, USA.

Berger, A. N., and Udell, L. F. (2006), A more complete conceptual framework for SME Finance, *Journal of Banking,* **30**.

Berger, A. N., Frame, W. S., and Miller, N. H. (2002), Credit Scoring and The Availability, Price And Risk Of Small Business Credit, US Federal Reserve System Working Papers, Washington, DC, USA.

Berger A. N., Klapper, L. F., and Udell, G. F. (2001), The Ability of Banks to Lend to Informationally Opaque Small Businesses, US Federal Reserve System Working Papers, Washington, DC, USA.

Berger, A. N., Miller, N. H., and Petersen, M. A. (2002), Does Function Follow Organizational Form? Evidence From the Lending Practices of Large and Small Banks, US National Bureau of Economic Research Working Papers, 8752, Cambridge, MA, USA.

Blochwitz, S., and Eigermann, J. (2000). Unternehmensbeurteilung durch Diskriminanzanalyse mit qualitativen Merkmalen, Zeitschrift fur betriebswirtschaftliche Forschung.

Bohn J. R. (2006), Structural Modeling in Practice, White Paper, Moody's KMV.

Boot A. W. (2000), Relationship Banking: What Do We Know? *Journal of Financial Intermediation,* **9**.

Boot, A. W., and Thakor, A. V. (2000), Can Relationship Banking Survive Competition? *The Journal of Finance,* **55**.

Brunetti, G., Coda, V., and Favotto, F. (1984), Analisi, previsioni, simulazioni economico-finanziarie d'impresa, Etas Libri.

Brunner, A., Krahnen, J. P., and Weber, M. (2000), Information Production in Credit Relationships: on the Role of Internal Ratings in Commercial Banking, Working Paper 10, Center for Financial Studies of University of Frankfurt, Germany.

Burroni, M., Quagliariello, M., Sabatini, E., and Tola, V. (2009), Dynamic Provisioning: Rationale, Functioning, and Prudential Treatment, *Questioni di Economia e Finanza,* **57**, Bank of Italy.

Buzzell, R. D. (2004), The PIMS Program of Strategy Research: A Retrospective Appraisal, *Journal of Business Research,* **57** (5).

Buzzell, R. D., and Gale, B. T. (1987), The PIMS principles, The Free Press.

Cangemi, B., De Servigny, A., and Friedman, C. (2003), Credit Risk Tracker for Private Firms, Technical Document, Standard & Poor's.

Committee of European Banking Supervisors (2005), Guidelines on The Implementation, Validation and Assessment of Advanced Measurement (AMA) and Internal Ratings Based (IRB) Approaches.

Christodoulakis, G., and Satchell, S. (2008), The Analytics of Risk Validation, Elsevier.

De Laurentis, G. (1993), Il rischio di credito, Egea.

De Laurentis, G. (2001), Rating interni e credit risk management, Bancaria Editrice.

De Laurentis, G. (Ed.) (2005), Strategy and Organization of Corporate Banking, Springer.

De Laurentis, G., and Gabbi, G. (2010), The Model Risk in Credit Risk Management Processes, in Model Risk Evaluation Handbook (eds G. N. Gregoriu, C. Hoppe, and C. S. Wehn), McGraw Hill.

De Laurentis, G., and Gandolfi, G. (Eds) (2008), Il gestore imprese, Bancaria Editrice.

De Laurentis, G., Saita, F., and Sironi, A. (Eds) (2004), Rating interni e controllo del rischio di credito, Bancaria Editrice.

De Lerma, M., Gabbi, G., and Matthias, M. (2007), CART Analysis of Qualitative Variables to Improve Credit Rating Processes, http://www.greta.it/credit/credit2006/poster/7_Gabbi_Matthias_DeLerma.pdf (accessed February 2010).

De Servigny, A., and Renault, O. (2004), Measuring and Managing Credit Risk, McGraw-Hill.

De Servigny, A., Varetto, F., Salinas, E. *et al.* (2004), Credit Risk Tracker Italy, Technical Documentation, www.standardandpoors.com (accessed February 2010).

DeYoung, R., Hunter, W. C., and Udell, G. F. (2003), The Past Present and Probable Future for Community Banks, Working Paper 14, Federal Reserve Bank of Chicago, USA.

Diamond, D. (1984), Financial Intermediation and Delegated Monitoring, *The Review of Economic Studies,* **51** (3).

Draghi, M. (2008), A System with More Rules, More Capital, Less Debt And More Transparency, Sixth Committee of the Italian Senate, Fact-finding Inquiry into the International Financial Crisis and its Effects on the Italian Economy, Rome, http://www.bancaditalia.it (accessed February 2010).

Draghi, M. (2009), Address by the Governor of the Bank of Italy, Annual Meeting of the Italian Banking Association, 8 July 2009, Rome, http://www.bancaditalia.it (accessed February 2010).

Dwyer, D. W., Kocagil, A. E., and Stein, R. M. (2004), Moody's KMV Riskcalc™ v3.1 Model, Technical Document, http://www.moodyskmv.com/research/files/wp/RiskCalc_v3_1_Model.pdf (accessed February 2010).

Ely, D. P., and Robinson, K. J. (2001), Consolidation, Technology and the Changing Structure of Banks' Small Business Lending, *Federal Reserve Bank of Dallas Economic and Financial Review,* First Quarter.

Engelmann, B., and Rauhmeier, R. (Eds) (2006), The Basel II Risk Parameters, Springer.

Fisher, R. A. (1936), The Use of Multiple Measurements in Taxonomic Problems, *Annals of Eugenics,* **7**.

Finger C. (2009a), IRC Comments, RiskMetrics Group, Research Monthly (February).

Finger C. (2009b), VAR is from Mars, Capital is from Venus, RiskMetrics Group, Research Monthly (April).

Frame, W. S., Srinivasan, A., and Woosley, L. (2001), The Effect of Credit Scoring on Small Business Lending, *Journal of Money Credit and Banking*, **33**.

Ganguin, B., and Bilardello, J. (2005), Foundamentals of Corporate Credit Analysis, McGraw-Hill.

Giri, N. C. (2004), Multivariate Statistical Analysis: Revised And Expanded, CRC Press.

Grassini, L. (2007), Corso di Statistica Aziendale, Appunti sull'analisi statistica dei bilanci, http://www.ds.unifi.it/grassini/laura/Pistoia1/indexEAPT2007_08.htm (accessed February 2010).

Golder, P. A., and Yeomans, K. A. (1982), The Guttman-Kaiser Criterion as a Predictor of the Number of Common Factors, *The Statistician*, **31** (3).

Gupton, G. M., Finger, C. C., and Bhatia, M. (1997), Credit Metrics, Technical Document, Working Paper, JP Morgan, http://www.riskmetrics.com/publications/techdocs/cmtdovv.html (accessed February 2010).

IASB (2009), Basis for Conclusions on Exposure Draft, Financial Instruments: Amortized Cost and Impairment, 6 November 2009.

Itô, K. (1951), On Stochastic Differential Equations, American Mathematical Society, 4.

Jackson, P., and Perraudin, W. (1999), Regulatory Implications of Credit Risk Modelling, Credit Risk Modelling and the Regulatory Implications Conference (June 1999), Bank of England and Financial Services Authority, London.

Landau, S., and Everitt, B. (2004), A handbook of statistical analyses using SPSS-PASW, CRC Press.

Loehlin, J. C. (2003), Latent Variable Models – An Introduction to Factor, Path, and Structural Equation Analysis, Lawrence Erlbaum Associates.

Lopez, J., and Saidenberg, M. (2000), Evaluating credit risk models, *Journal of Banking and Finance*, **24**.

Lyn, T. (2009), Consumer Credit Models – Pricing, Profit and Portfolios, Oxford Scholarship Online.

Maino, R., and Masera, R. (2003), Medium Sized Firm and Local Productive Systems in a Basel 2 Perspective, in Industrial Districts and Firms: The Challenge of Globalization, Modena University, Italy, Proceedings, http://www.economia.unimore.it/convegni_seminari/CG_sept03/papers.html (accessed February 2010).

Maino, R., and Masera, R. (2005), Impresa, finanza, mercato. La gestione integrata del rischio, EGEA.

Masera, R. (2001) *Il Rischio e le Banche*, Edizioni Il Sole 24 Ore, Milano.

Masera, R. (2005), Rischio, Banche, Imprese, i nuovi standard di Basilea, Edizioni Il Sole 24 Ore.

Masera, R., and Mazzoni, G. (2006), Una nota sulle attività di Risk e Capital Management di un intermediario bancario, Ente Luigi Einaudi, Quaderni, **62**.

Merton, R. (1974), On the pricing of Corporate Debt: the Risk Structure of Interest Rates, *Journal of Finance*, **29**.

Modigliani, F., and Miller, M. H. (1958), The Cost of Capital, Corporation Finance and the Theory of Investment, *American Economic Review*, **48**.

Moody's Investor Services (2000), Benchmarking Quantitative Default Risk Models: a Validation Methodology (March).

Moody's Investor Service (2007), Bank Loan Recoveries and the Role that Covenants Play: What Really Matters? Special Comment (July).

Moody's Investor Service (2008), Corporate Default and Recovery Rates 1920–2007 (February).

Nixon, R. (2006), Study Predicts Foreclosure for 1 in 5 Subprime Loans, *NY Times* (20 December 2006).

OeNB and FMA (2004), Rating Models and Validation, Oesterreichische Nationalbank and Austrian Financial Market Authority.

Petersen, M. A., and Rajan, R. G. (1994), The Benefits of Lending Relationships: Evidence From Small Business Data, *Journal of Finance,* **49**.

Petersen, M. A., and Rajan, R. G. (2002), Does Distance Still Matter? The Information Revolution in Small Business Lending, *Journal of Finance,* **57** (6).

Pluto, K., and Tasche, D. (2004) Estimating Probabilities of Default on Low Default Portfolios, Deutsche Bundesbank Publication (December).

Porter, M. (1980), Competitive Strategy, Free Press.

Porter, M. (1985), Competitive Advantage: Creating and Sustaining Superior Performance, Free Press.

Rajan, R. G. (1992), Insiders and Outsiders: the Choice Between Relationship and Arms Length Debt, *Journal of Finance,* **47**.

Resti, A., and Sironi, A. (2007), Risk Management and Shareholders' Value In Banking, John Wiley & Sons Ltd.

Saita, F. (2007), Value at risk and bank capital management, Elsevier.

Schwizer, P. (2005), Organizational Structures, in Strategy and Organization of Corporate Banking (Ed. G. De Laurentis), Springer.

Sharpe, W. (1964), Capital Asset Prices: a theory Of Market Equilibrium under Conditions of Risk, *Journal of Finance,* **19**.

Sobehart, J. R., Keenan, S. C., and Stein, R. M. (2000), Validation Methodologies For Default Risk Models, *Algo Research Quarterly,* **4** (1/2) (March/June).

Standard & Poor's (1998), Corporate Ratings Criteria, http://www.standardandpoors.com.

Standard & Poor's (2008), Corporate Ratings Criteria, http://www.standardandpoors.com.

Standard & Poor's (2009), Default, Transition, and Recovery: 2008 Annual Global Corporate Default Study And Rating Transitions.

Standard & Poor's (2009a), Annual Global Corporate Default Study and Rating Transitions, http://www.standardandpoors.com.

Standard & Poor's (2009b), Global Structured Finance Default and Transition Study 1978–2008: Credit Quality of Global Structured Securities Fell Sharply in 2008 Amid Capital Market Turmoil, http://www.standardandpoors.com.

Standard & Poor's (2009c), Guide To Credit Rating Essentials, 21 August 2009, http://www.standardandpoors.com.

Steeb, W. H. (2008), The Nonlinear Workbook: Chaos, Fractals, Neural Networks, Genetic Algorithms, Gene Expression Programming, Support Vector Machine, Wavelets, Hidden Markov Models, Fuzzy Logic with C++, Java and Symbolic C++ Programs: 4th edition, World Scientific Publishing.

Stevens, J. (2002), Applied Multivariate Statistics for the Social Sciences, Lawrence Erlbaum Associates.

Tan, P.-N., Steinbach, M., and Kumar, V. (2006) Introduction to Data Mining, Addison-Wesley.

Tarashev, N. A. (2005), An Empirical Evaluation of Structural Credit Risk Models, Working Papers No. 179, BIS Monetary and Economic Department, Basel, Switzerland.

Thompson, M., and Krull, S. (2009), In the S&P 1500 Investment-Grade Stocks Offer Higher Returns over the Long Term, Standard and Poor's Market Credit and Risk Strategies (June), http://www.standardandpoors.com.

Thurstone, L. L. (1947), Multiple Factor Analysis, University of Chicago Press, Chicago.

Treacy, W. F., and Carey, M. S. (1998), Credit Risk Rating at Large U.S. Banks, US Federal Reserve Bulletin (November).

Treacy, W. F., and Carey, M. S. (2000), Credit Risk Rating Systems at Large U.S. Banks, Journal of Banking and Finance, 24.

Tukey, J. W. (1977), Exploratory Data Analysis, Addison-Wesley.

Udell, G. F. (1989), Loan Quality Commercial Loan Review and Loan Officer Contracting, Journal of Banking and Finance, 13.

Vasicek, O. A. (1984), Credit Valuation, White Paper, Moody's KMV (March).

Wehrspohn, U. (2004), Optimal Simultaneous Validation Tests of Default Probabilities Dependencies and Credit Risk Models, http://ssrn.com/abstract=591961 (accessed February 2010).

Wilcox, J. W. (1971), A Gambler's Ruin Prediction of Business Failure Using Accounting Data, Sloan Management Review, 12 (3).

Index

Accuracy ratio AR, 152, 203, 204, 206
Adverse selection, 289
Agencies' ratings, 31, 301
Altman, 7, 21, 33, 39, 40, 43, 44
Annualized default rate, 29
ANOVA, 145, 149, 150, 166
Artificial intelligence, 77–79
Assets value volatility, 34, 36
Asymmetry of a distribution, 134

Backward chaining, 79
Basel II, 1–3, 97, 100, 229, 234, 237–239, 244–246, 250, 257, 258, 276, 289, 291–293, 298, 300, 305, 309, 310, 312, 313
Behavioral data, 58, 59, 86, 93, 96, 97, 207, 290, 300, 303, 304, 309
Binomial test, 251
Bivariate analyses, 94, 95, 96, 116, 177, 198, 210, 216
Black box, 84, 86, 244, 310
Black Scholes Merton formula, 34, 35
Bootstrapping method, 114, 187, 250
Bottom up approach, 41, 306, 307
Boxplot, 142, 143, 144, 266–267, 278
Business risk, 20, 24, 25, 35, 126, 299

Calibration, 37, 40, 47, 48, 52, 53, 58, 96, 100, 115, 229–231, 235, 239, 242, 243, 247–252, 256, 275, 289, 298, 300, 305, 314
CAMELS, 21
Canonical correlation analysis, 61, 72
Canonical discriminant function coefficients, 195, 196, 210, 273, 283
Case study, 97, 257
Categorical covariates, 212
Central tendency, 51, 62, 95, 108, 117, 132, 140, 198, 248, 249
Centroids, 45–47, 195, 198, 202, 207, 283
Chi-square test, 151, 214, 230, 250, 252
Cluster analysis, 60–62, 77
Co-linearity, 47
Commercial function, 294
Conditional probability, 49
Confidence interval, 18, 57, 132, 135, 154, 207, 230, 231, 235, 256, 275, 283
Consumer loan, 48
Contingency tables, 151, 202, 203, 208, 250, 259
Continuous annual default, 29
Core earnings methodology, 26
Correlation, 11, 12, 160–162, 248, 270
Coverage ratio, 24, 80
Credit derivatives, 1, 13
Credit function, 294, 308

Developing, Validating and Using Internal Ratings: Methodologies and Case Studies Giacomo De Laurentis, Renato Maino and Luca Molteni © 2010 John Wiley & Sons, Ltd